顶级设计空间
TOP DESIGN SPACE
《顶级设计空间》编委会编

情调餐厅
ROMANTIC RESTAURANT

中式餐厅 & 西式餐厅
日式餐厅 & 酒吧
CHINESE & WESTERN-STYLE & JAPANESE RESTAURANT & BARS

（第二版）

中国林业出版社

情调餐厅：汉英对照 / 《顶级设计空间》编委会编
. -- 2版. -- 北京：中国林业出版社, 2013
（顶级设计空间）

ISBN 978-7-5038-7272-3

Ⅰ. ①情… Ⅱ. ①顶… Ⅲ. ①餐馆－室内装饰设计－图集 Ⅳ. ①TU247.3-64

中国版本图书馆CIP数据核字(2013)第274868号

《顶级设计空间》编委会编

主编：张青萍
编委：孔新民、贾陈陈、许科、李钢、吴韵、竺智、曾丽娴

责任编辑：纪亮
英文翻译：董君、梅建平、牛晓霆、万毅、赵强

出版：中国林业出版社
（100009 北京西城区德内大街刘海胡同 7 号）
网址：http://lycb.forestry.gov.cn/
E-mail：cfphz@public.bta.net.cn
电话：（010）83143581
发行：新华书店
印刷：北京卡乐富印刷有限公司
版次：2016年4月第2版
印次：2016年4月第1次
开本：230mm × 300mm
印张：32
字数：360千字
定价：199.00元

设计进行时

中国改革开放30年，室内设计行业行进到今天，也已经有了无计其数的变化、发展和积累，30年的思考、30年的实践、30年的进步，也造就了这30年的成绩。

我们好似在进行着一场接力赛，祖先把中华民族灿烂的文化一代代地传承到21世纪，我们有责任将这份优秀的遗产倍加珍藏以传给后代。我们所面临的挑战是拿什么当代的室内文化馈赠后人？但其实这30年中由于电子技术的普及和信息的迅速传播和交换，设计已出现国际化、同一化的倾向，与此同时引起了传统性、地域性和个性差异的不断丧失，又有由于社会追求物质与功能价值的同时造成对精神和文化价值的忽视，我们已找不到回头的路。但不管历史结论会如何，我们这代人是努力的、勤奋的，是不断地用自己的智慧为中国室内设计行业进步奉献着的。

总的看来，21世纪的室内设计发展有以下倾向和趋势：

倡导绿色设计

人类起源于自然，其间虽曾摆脱过自然，但最终还是要以全新的面目去回归于自然。如此轮回恰恰历史地、辩证地道出了人与自然关系的变化。如今的人们，特别是生活在大城市里的人离大自然已越来越远了，于是人们特别希望在室内再现一些大自然的情景，以求得哪怕是暂时地、局部地享受。作为设计师一方面尽可能地创造出生态环境，让人们最大限度地接近自然，另一方面须有环保意识，努力去提高设计中的健康因素，以满足人们在生理和心理上的需要。

室内设计中的健康设计充分利用自然或仿自然的因素，为人们提供生活舒适的空间。室内的色彩、照明及功能空间的弹性分割，都应该在满足其基本功能的基础上，尽可能充分利用自然能源。尤其是提倡对绿色装饰材料和绿色照明材料的运用，同时注重社会心理学的研究。绿色设计本着以下几个原则：

①设计上使用最少的材料，少浪费应节约的资源、能源，力戒奢靡；②尽量多采用污染少，环保性能强，安全可靠的材料；③符合人体工程学的要求，讲求空间上比例与尺度，避免使人感到压抑或繁琐；④设计应以人为本，满足人的本质需要。

运用高科技手段

科技进步影响着人类生活的方方面面，现代科技的发展为人类的衣食住行提供了很多方便，它可以使居住环境更加符合人们的意愿。因此，在设计中提高科技含量，创造高效率、高功能、高质量室内生活环境的要求已愈加鲜明。

在本世纪，楼宇的智能化将逐步实现。建筑智能化就是将结构、设备、服务运营及相互关系进行全面综合配置，从而达到最佳的组合程度，使建筑具有高效率、高功能和舒适性。单从自动化来看，就是实现建筑设备自动化、办公自动化和通讯自动化。建筑智能化并非仅仅针对大楼宇，还会迅速走向寻常百姓家。住宅中自动防盗报警，自动调温、调湿、自动除尘、调节灯光亮度、自动控制炊事用具等，如今也已成为事实。

注重设计的系统性

在这个日新月异、急剧变化的时代，系统地看待问题和解决问题是当代人的特质。站在现代与历史之间的人们既希望从传统中找回精神的家园，以弥补快速发展带来的心理失落与不安；同时又满怀着激情和野心试图运用当代技术和审美重新诠释历史，使之适应现代生活。系统的眼光可使我们将面临的室内设计浅层次问题渗透到更深的层次方面加以科学、综合地解决。这样的设计系统性包括环境学、生态学、经济学、系统论、方法论、控制论、统筹学、管理学及有关室内设计方面的政策法规、标准规范等方面的内容。

室内设计系统是指应用系统的观点和方法，将室内设计的内容、要素，相关的领域和环节，以及室内设计的程序予以统筹而形成的一个框架体系。从与其相关部分的关系和进行的程序来分析，可理解为有横向设计系统和纵向设计系统两个方面。横向系统设计表现为在设计过程中所涉及到的如生理学、心理学、行为科学、人体工程学、材料学、声学、光学、经济学等诸多因素；纵向系统设计表现为对设计实现过程中所有历程的考虑。概括而言，横向系统设计强调相关与联系；纵向系统设计强调过程与变化。

无论将来的室内设计如何发展，它都必然在一个更光泛、更全面的系统里科学地伸展，设计师也必将持有这种科学的态度和掌握这类理性的方法而从事设计。

本系列图书刊登的是近年来一些顶级的设计作品，它们或多或少的反映的是当下室内设计师的思考和当前技术背景下的实践。30年告了一个段落，下一个30年又将开始，我们已走上新的征程，设计永远是进行时。愿这本套书的出版能得到业界的认可和赞扬！

南京林业大学 风景园林学院副院长、教授
id+c《室内设计与装修》杂志主编　张青萍
2010.3.1草于南京

中式餐厅

日式餐厅

Chinese Restaurant

Western-style Restaurant

Japanese Restaurant

Bars

FU LU RESTAURANT

北京福鹿名肴会

01

【坐落地点】北京CBD商圈旺座中心；【面积】餐厅1 800 m²；厨房200 m²
【设计】耿治国；【设计单位】飞形设计事业有限公司
【主要建材】帝王石、宇宙金麻、黑镜、木花格、亚克力门洞、木皮壁纸、斗拱艺术品、艺术砖
【摄影】Nobuko

本案中的餐厅利用中国庭园的借景、框景，在有限的空间内创造出无限的想象。餐厅区域划分明朗，共由7大部分组成：两间超豪华VIP包房、开放主题用餐区、酒吧演艺厅区域、“冰山火海”玻璃区、日式铁板烧区域、现代江南风情包房区和西餐区。

一进大门，迎宾台对面就有一处点睛之笔：一个硕大如斗拱形状的“绿如意”装饰。出于私密性的考虑，餐厅的两个超级豪华的VIP包房区设在了空间的最左侧。穿过中式古典风格的长廊与大厅来到包房，门上花瓶形状的玻璃设计使人眼前一亮。隔着门中间红色半透明玻璃，衬托出VIP包房一种朦胧的美感。用餐区主位的坐凳明显与其他座位不同，体现客人的尊贵身份，从中可以看出餐厅符合中国传统礼仪的风格。

餐厅主题用餐区，可以称得上是结合了餐饮与视觉的现代光影庭园。开放用餐区被分成23个小区域，并且每一个区域都被装饰成形态各异的“小包间”，像一个个“半透明”的可爱“小屋”，而且每个“小屋”都被赋予一个特点和一个名字，而且在强调私密性的同时，让置身其中的人们感觉宽敞和通透。

此外，日式铁板烧区、“冰山火海”玻璃区、现代江南风情包房区及西餐区等各具特色，可以让客人根据自己的喜好，在花样繁复的选择中找到自己的最爱。

This case, the restaurant through the use of Chinese garden scene, box King, in the limited space to create a limitless imagination. Restaurants zoning uncertain, comprises seven main components: two ultra-luxury VIP rooms, opening up the theme dining area, bar auditorium regions, "iceberg sea of fire" glass area, Japanese-style teppanyaki area, modern Jiangnan style private dining room district and Western district.

Entry out there is a huge "green Ruyi" decoration. From the hidden point of view, the restaurant's two luxurious VIP private dining rooms are designed in an empty area to the left.Chinese classical style through the long corridors and halls, door glass, vase-shaped design makes shines. Next to the middle of red translucent glass, bring out the VIP private room of a hazy beauty. Theme of the stool dining area is obviously different from the other, reflecting the distinguished guests as can be seen from dining etiquette in line with traditional Chinese style.

Restaurant theme dining area, could be described as a combination of the modern garden dining and visual. An open dining area is divided into 23 small areas and each area had been decorated as a different "packet rooms", like an a "translucent" and "hut" and each "house" is given a theme, but also stressed that, while privacy so that people feel spacious and transparent.

In addition, the Japanese-style iron barbecue area, "tip of a sea of fire" glass area, modern Jiangnan district and Western district, such as private rooms styles vary, guests according to their own preferences, choose to find their own favorite.

↑福鹿名肴会入口 / Fu Lu Restaurant entrance.

↑VIP包间休息区 / VIP rooms rest area.
←硕大如斗拱形状的“绿如意”装饰在入口进门处 / Enter the gate at the entrance, huge arched "green Ruyi" decoration.
↓平面图 / Plan

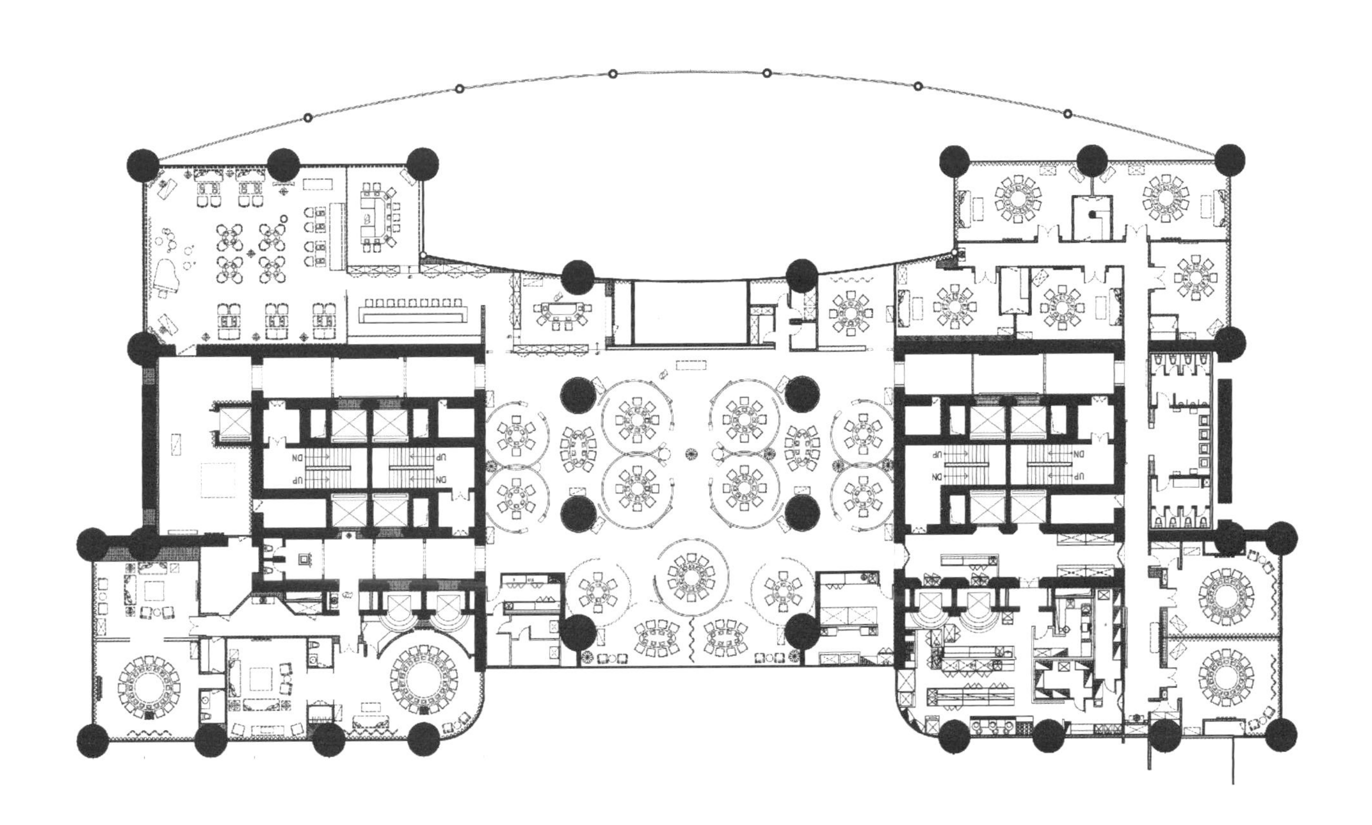

↑日式铁板烧区 / Japanese-style Teppanyaki area.
→主题用餐区的小包间 / Topics of packets between the dining area.
↓西式水晶灯结合中式缎面靠背椅，营造出幽雅的意境 / Western-style chandeliers with Chinese-style chairs, creating a mood of elegant.

↑卫生间洗手盆的支架也采用的如意的形体 / Fu Lu Restaurant entranceBathroom wash basin shape on the use of Ruyi.
←VIP包间就餐区 / VIP rooms dining area.
↓卫生间，贴面为手工烧制釉面的瓷砖 / Veneer bathroom for hand-fired tiles；↓庭园的借景/ Chinese garden location.

SOUTH BEAUTY AT SUN SQUARE

俏江南阳光广场店

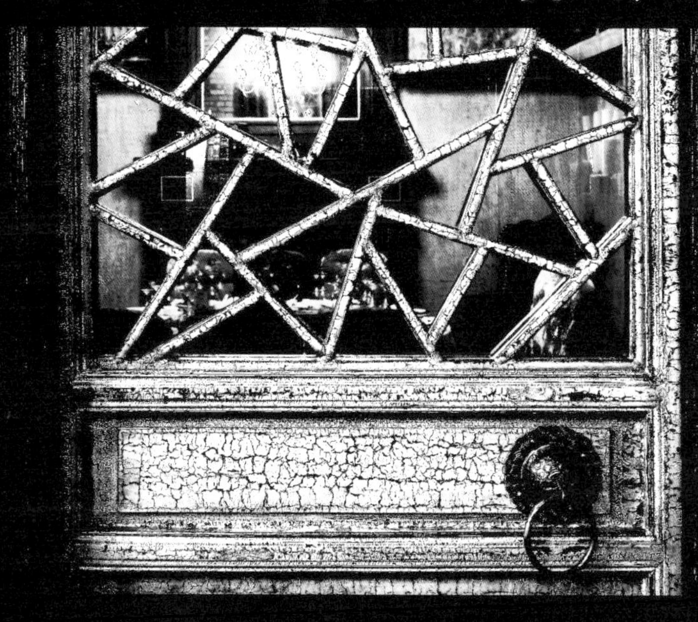
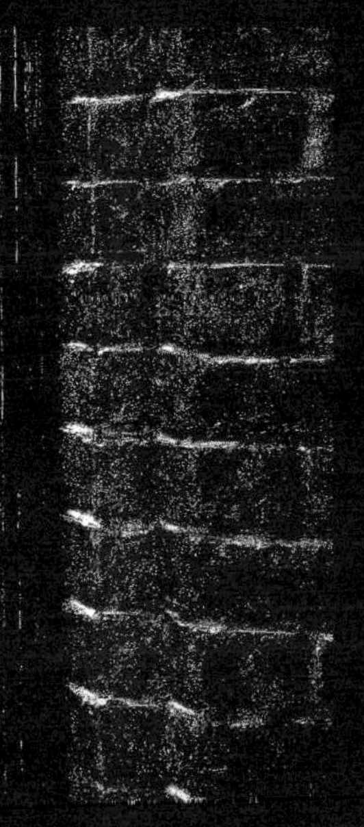

02

【坐落地点】北京亚运村安立路68号；【面积】1 800 m²

【设计】田军；【参与设计】韩玉权、全宏博

【设计单位】北京瑞普&田军工作室

【摄影】贾方

餐厅处于破败与奢华之间，传统与现代之间，东方与西方之间，两者兼有，却不偏不倚。设计师把它比作一个“仓库”，破败——正是设计师灵感的来源。同时，在设计中运用了众多元素与手法，它们被设计师精心的安排组织起来。矛盾对比与冲突，成为该餐厅室内体验的焦点，令人难忘。

餐厅设计酷感十足，独特的就餐环境在北京可堪称独树一帜，以黑、白、红为主色的整体，以钢、玻璃、红砖为主的材料，所有传统的元素都得到颠覆性的现代化包装，透明的吊桥，玻璃的月亮门，闪着灯光的蒙古包包间，线条简约的藤制坐椅，点缀着中国味十足的翠竹、珠帘、民乐，呈现出东方和西方的融合。

餐厅中采用红砖、古董地板、旧的花格门，甚至用上百年的旧木头做隔墙，来强调旧的氛围，散乱、随意、高大、破败……餐厅具有种种旧仓库的特性。同时，铜制的长短水晶吊灯、色彩艳丽的太湖石、镀金台灯、大红色真皮沙发……又打造了一种截然不同的奢华品质。破败与奢华在这里交织、冲突，让人不解，诱发了对此设计的进一步阅读。

不仅如此，设计师将中式与欧式的物品并置，又为餐厅的室内设计增添了一个层次。为避免杂乱，设计师巧妙地将中与西的物品运用在不同的层面。中式的红蜡烛、花格门、冰裂纹玻璃、水墨字画、太湖石用于装饰；而欧式的沙发、灯具等具有实际的使用功能。对于中式的精神和西式的物质相互区分，也在暗示着食客：在享受奢华的物质生活时，也不要忘记享受精神生活。

Restaurant in the case between the old and the luxury is a fusion of traditional and modern culture, but also the integration of Eastern and Western culture. Designers liken it to a "warehouse", old - it is a source of inspiration for designers. Meanwhile, in the design of the use of many elements and practices, they are designers carefully arrange organized. Experience, the focus of the restaurant interior is contradictory comparison, contrast is unforgettable.

Restaurant design is very personal and unique environment in Beijing might be the only to black, white, and red color options to steel, glass, brick-based materials, all the traditional elements have been fundamental changes and Packaging, a transparent bridge, glass, moon gate, flashing lights Mongolian-style private rooms, a simply-rattan chair, full of Chinese-style bamboo bead curtain, folk music, show a blend of East and West.

Restaurant use of red brick, old flooring, old plaid door, or even spend a hundred years to do the old wooden walls, to emphasize the atmosphere of the old, messy, random, tall, old ... ... restaurant has a variety of characteristics of an old warehouse. Meanwhile, the copper crystal chandeliers, colorful stones, gold-plated table lamp, red leather sofa ... ... also created a different kind of luxury quality. A better understanding of the design of the case are: worn with the luxury of space inside the convergence.

Moreover, Chinese and European designer items will be put together for the restaurant's interior design adds a level. Designer Chinese and Western-style items skillfully placed in a different space, to avoid confusion of space. Chinese candles, lattice doors, with a pattern of glass, Chinese calligraphy, stone, etc. to decorate the environment; while the European sofa, lamps, etc. with the actual use function.

↑花格窗被涂上鲜艳的红色，红蜡烛被设计师做的粗壮高大 / Lattice windows have been painted bright red, red candles are tall and sturdy design to do.

↑餐厅中采用红砖、古董地板、旧的花格门 / Restaurants use red brick, old floor, the old lattice door.

↓奢华的坐椅与破败的红砖引发冲突，却又统一在红色色系中 / In the red, the luxury seats and the contrast between the dilapidated red-brick.

↑水晶吊灯、镀金台灯、红色真皮沙发营造了一种奢华品质 / Crystal chandeliers, gold-plated table lamp, red leather sofa to create a luxurious quality.
↓平面图 / Plan

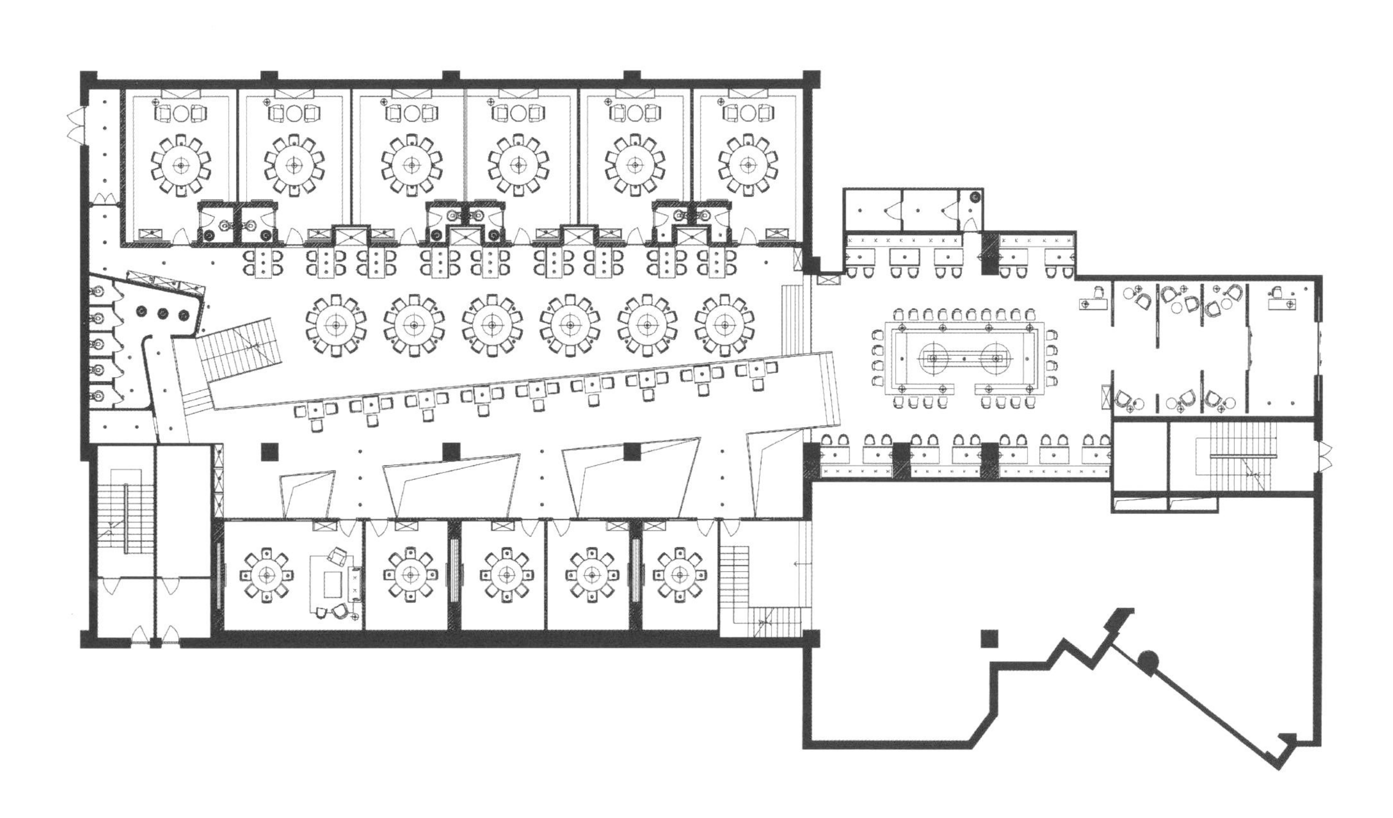

↑夹层楼板完全采用玻璃 / Floor full of glass.

↓天花板丝毫不加修饰，管道、梁都暴露在外 / Without the slightest modification of the ceiling, pipes, beams were exposed.

↑若干个包间采用纯欧式的装饰，与包间外形成对比，别有洞天 / A number of European-style rooms decorated with pure, in contrast with the outside rooms.
↓鲜艳的蓝色太湖石成了“窗”外的景色 / Bright blue Taihu stone become a "window" outside view.

↑夹层楼板完全采用玻璃 / Floor full of glass.
→包间采用纯欧式的装饰，与包间外形成对比，别有洞天 / European-style rooms decorated with pure, in contrast with the outside rooms.
↓新旧材质的对比：红砖、花格门与玻璃、工字钢 / Comparison of old and new materials: red brick, lattice doors and glass, steel frame.

WHAMPOA CLUB

北京黄浦会餐厅

03

【坐落地点】北京金融街甲23号
【面积】2 080 m²
【设计】如恩设计研究室
【摄影】Derryck Menere

黄浦会餐厅的首层为酒吧长廊和贵宾包厢，地下一层为主就餐区、半包厢区域。设计需集中解决的问题：如何在一个传统中国庭院的建筑里把现代用餐体验和传统中国饮食很好的结合起来。

进入餐厅，纯白色的走廊营造出纯净的空间演绎着中式建筑的细节而非其他晦涩的色彩来装点。白色的长廊让即将要走入其他变化空间的人们以充分的放松。灯光流过"断裂"的庭院，回旋在餐厅那数以万计的悬吊的不锈钢片之间。这些不锈钢片既将人们各自的形象映入，更以一个提醒者的身份，让人们通过它们从上方瞥见整个院落。这个下沉的庭院被一股细流围住，与上方的水池相映成趣，同时，乌木色的屏风重构了庭院周围门厅和房间的空间。屏风之后是一个半私人的用餐空间，提供了一个回顾庭院风景的机会。

贵宾包厢的色彩各不相同，由特别设计定制的漆器屏风构架而成。其中最能体现传统中国文化特质的当属最北端的包厢，最为吉祥，最为精致繁复。楼梯象征并宣扬了中式菜肴的准备过程。拾级而下，人们一步一步的体验了中餐的烹饪方式所独有的慢工细活，而当人们走完楼梯后，就可以入座来享用食物本身的美味了。

This case the first floor gallery and VIP boxes for the bar, underground layer of the main dining area, semi-box area. Need to focus on solving the design problem: how a traditional Chinese courtyard building where modern dining experience and traditional Chinese food very good together.

When people entered the restaurant, restaurant in the corridors of pure white Chinese-style architectural details of the show.Changes in the white gallery space so that access to the people to relax fully. Light flows through the garden, irradiation in a restaurant that tens of thousands of suspended between stainless steel sheet. These stainless steel sheet illuminated the image of their own people, let people through their time, can see the entire courtyard. The courtyard was surrounded by a trickle, with the top of the pool echoes the same time, reconstruction of the courtyard wall around the hall and room space. Screen followed by a semi-private dining space, creating a garden landscape recalls the atmosphere.

VIP box color varies from a specially crafted lacquer screen composition. One of the most embodies traditional Chinese culture among the most northern end of the box, good luck, and refined. And the promotion of the stairs symbolizes the process of preparing Chinese dishes. Through the staircase, one step by step, the experience of the Chinese cuisine cooking methods, and when one has completed the stairs, you can come to enjoy the delicious food seating.

↑白色的入口长廊 / White entrance corridor.

↑华丽的主就餐区，让人想起古老建筑前世的辉煌 / Magnificent main dining area, reminiscent of the old building's former glory.
←中式四合院与西式建构的对比 / Chinese courtyard houses contrast with the Western construct.
↓一层平面图 / Floor plan.

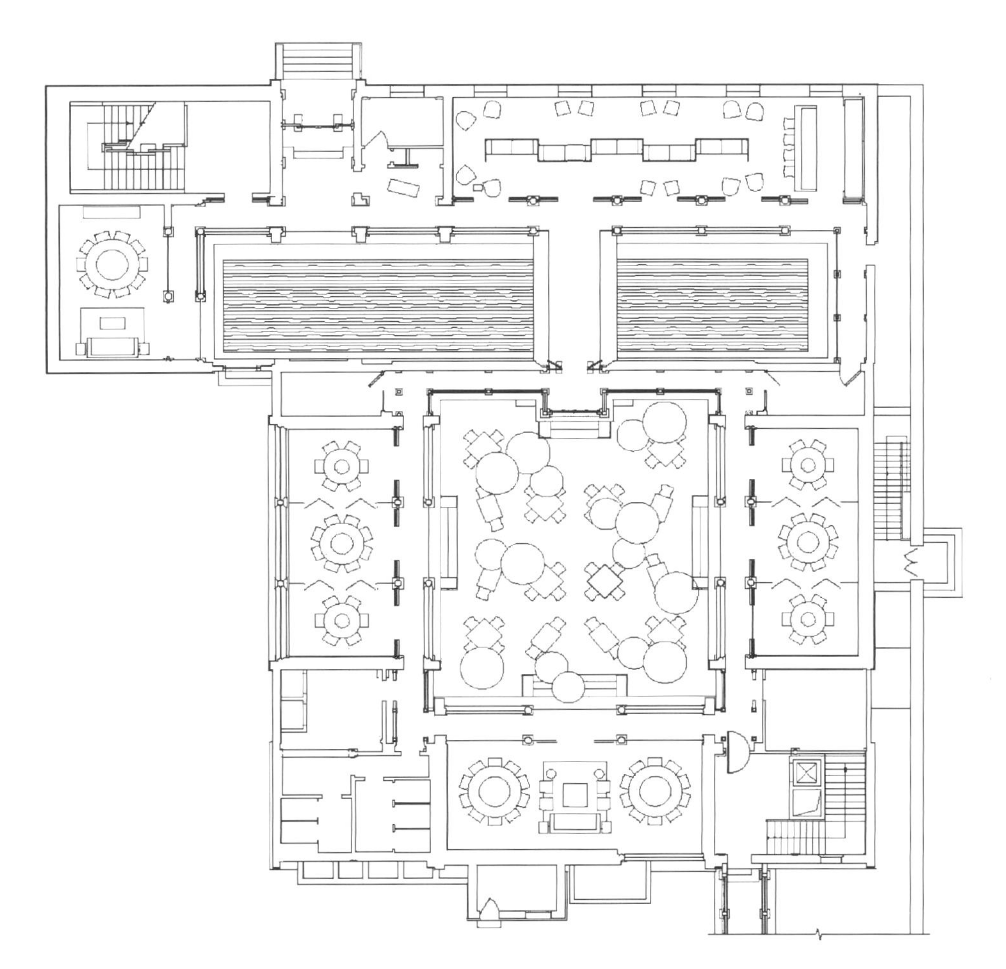

↑中式的隔扇和西式的幔帘共处一处 / Chinese windows and Western-style mix of gauze；↑楼梯 / Stairs
←传统的鸟笼和现代的家具 / Traditional cage and modern furniture.
↓地下一层平面图 / Basement floor plan.

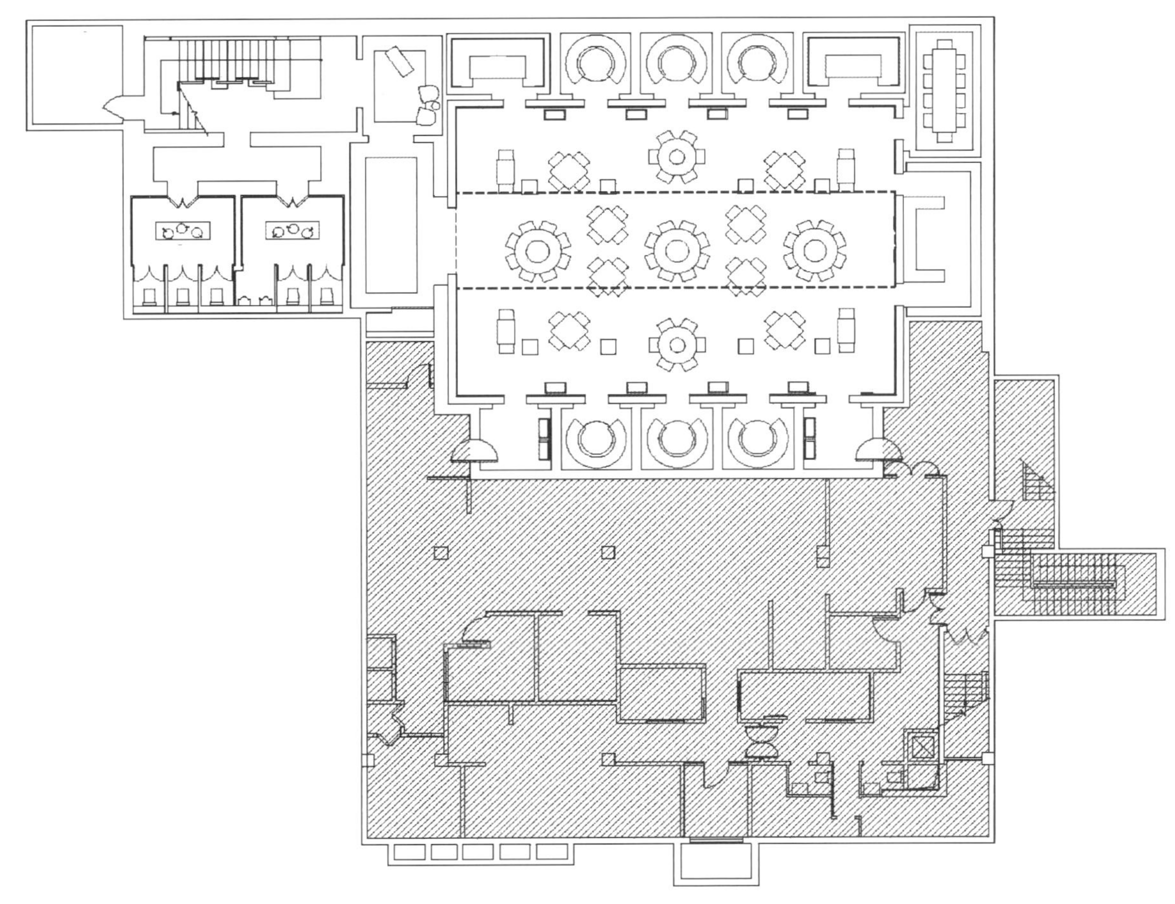

↑贵宾包厢，用中国传统元素展现现代餐饮 / VIP boxes with traditional Chinese elements to show a modern restaurant.
←酒吧区 / Bar area.
↓特色洗手间 / Toilet

JIN SHA YUAN NIAN

成都“金沙元年”食府

04

【坐落地点】成都市“金沙遗址”公园 ；【面积】6 000 m^2
【设计】杨樵
【主要用材】 法国木纹石、美国枫木、皮雕
【摄影】贾方

原建筑群本身就是一个开放式的院落，极富远古的感染力。在设计时不能把室内和室外切割开，二者必须是一个整体。后来，我们用木构架作了一个较大的延伸体，既增加了建筑体的灵动性，又将原来分离的茅屋有机连接起来。通过巨大的岷江石、粗大的树木、几组老木头的构架，以及古朴的石雕，在花岗石石条和草坪的间构中，简洁而真实地反映出古代院落的生活场景。

金沙遗址里最有美感的立体器物就是象征“天圆地方”的玉琮，因为它特别能表达古人的一种情感诉求。通过提炼，我们把“天圆地方”这种最朴素的古人哲学思想贯穿到了整个食府。比如：庭院入口就是一个玉琮放大的抽象造型，所有门套、吧台的酒架和吧柜所有的餐椅都可以看到天圆地方的形状，公共区域的灯具，水景的器具，都充满了天圆地方的概念。

院落象征着古人的物质述求，天圆地方象征着古人的精神述求。从开业后极好的经营状况，以及对室内外设计的良好口碑来看，整体设计最后都达到了最初的想法，即在沉稳和低调中彰显一种复古的奢华！

The original building complex itself is an open courtyard, very ancient infectious. In the design of indoor and outdoor can not be cut open, the two must be a whole. Later, we made a large wooden frame with an extension of the body so as to enhance the building of Smart physical nature of the original separation of the house again to connect. Minjiang River through the use of a huge stone, thick trees, wood framework, and the ancient stone, granite stone in the middle section and lawn, simple and true reflection of the lives of the ancient courtyard scenes.

Jinsha ruins most aesthetic objects is a symbol of "Heaven and Earth," the jade, because it specifically to express an emotion to pray for the ancients. Through refining, we have the "Heaven and Earth" This simple philosophy of the ancients through to the entire restaurant. For example: the courtyard entrance is a larger form of jade, all the door frames, cabinets bar of wine racks and bar, all the dining chair can be seen, "side and round" shape, the public area lighting, water features of the apparatus, are filled with The "square and the circle" concept.

The material symbol of ancient courtyard to pray, "side and round" symbolizes the spirit of the ancients to pray. From the opening after the excellent operating conditions, as well as a good assessment of indoor and outdoor design point of view, the overall design eventually reach the initial thought is reflected in the steady and low-key kind of retro luxury!

↑夜幕下的食府，那么安静幽雅 / Night under the restaurants, then quietly elegant.

↑粗犷的石头、木质的框体，再加上丝绸的沙发靠垫，刚与柔的对比 / Rough stone, wood box body and silk, in contrast to the sofa cushion.
→突出的层高抹杀了斜屋顶本应带来的局促感 / Prominent storey changes brought about by cramped sense of sloping roofs.
↓平面图 / Plan

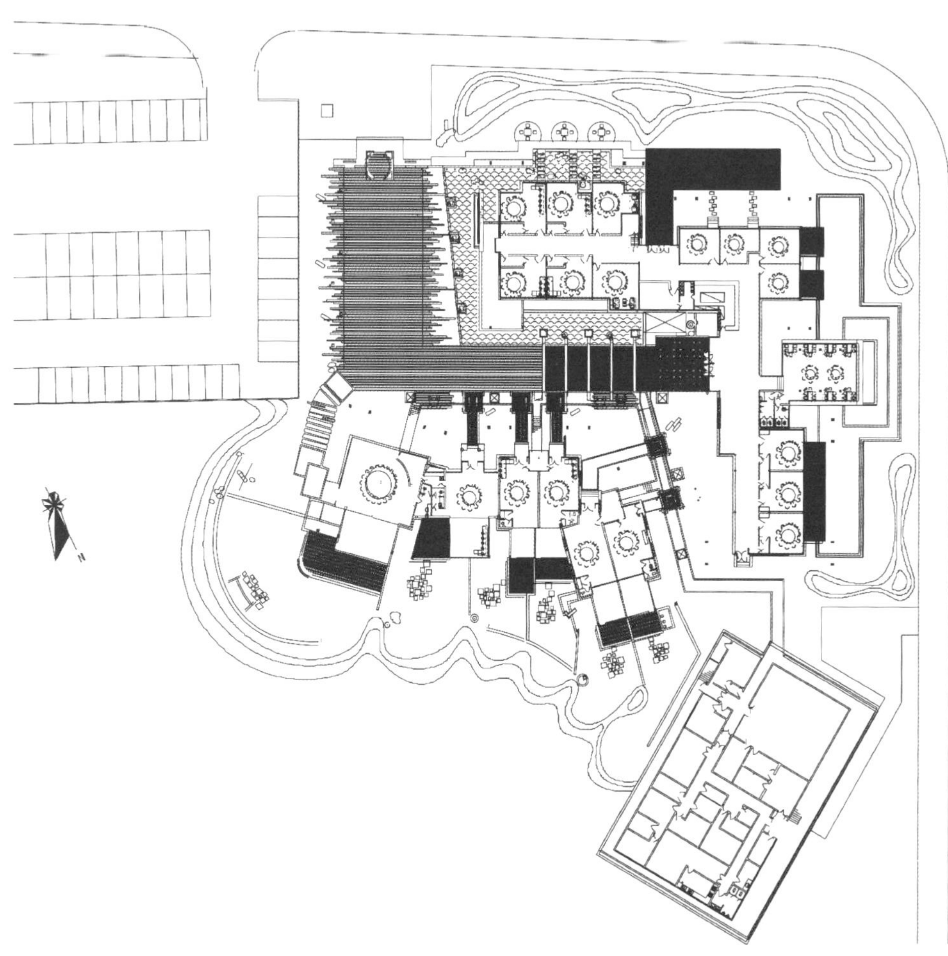

↑入口处成列的各种工艺品 / Columns at the entrance into a variety of crafts.
→突出的层高抹杀了斜屋顶本应带来的局促感 / Prominent storey changes brought about by cramped sense of sloping roofs.
↓豪华包厢 / Luxury boxes.

↑喜庆的包厢 / Festive box.

←↓包厢根据不同需要或开或合 / Box according to the different needs of separate or combined.

SOUTH BEAUTY

Subu俏江南北京店

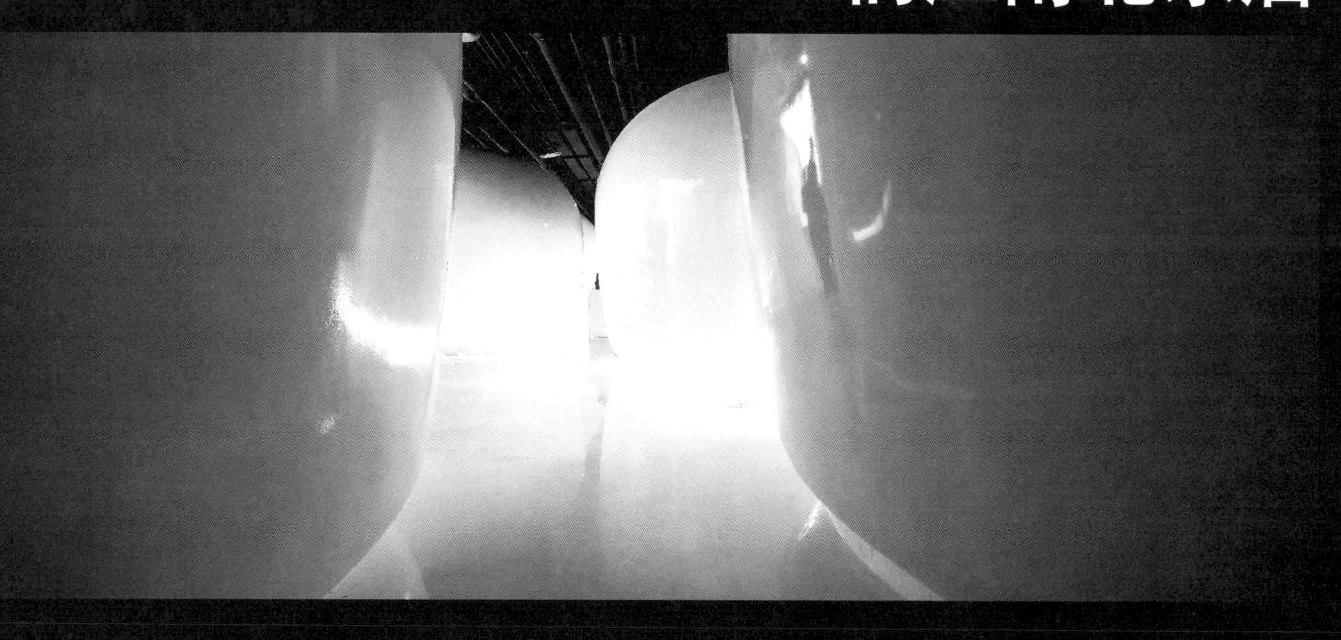

05

【坐落地点】北京金融街购物中心四层
【面积】1 400 m^2
【设计】Johannes Torpe；【设计公司】Johannes Torpe工作室
【摄影】贾方

本店位于金融街购物中心四层，店分为两部分，商场天庭下的开放区和右侧的封闭区。餐厅走的是国际化和年轻时尚化的路线，超前卫的外部设计，大胆诱人的色彩配置，让用餐者仿佛置身在宇宙飞船内。

整个餐厅采用了大量的玻璃材料，它在半封闭的状态下将室内外联系起来，对顾客的视线没有丝毫阻挡。但玻璃所牵引的视觉上的联系毕竟是感性的，为了进一步让内外空间更加理性的贯穿在一起，将人的情感和活动融入进去，设计师联想到了桥，由此引申出了拱廊的想法——在玻璃的基础上，搭建拱廊的结构，真正创造了如太空舱一般的外部造型。

富于变化的色彩激发了用餐空间的无限活力，完美灵动的绚丽颜色肆无忌惮的捕捉着人们的视线，点亮了整个内部环境。SUBU不仅仅是个循规蹈矩的餐厅，更是一种全新的时尚生活理念，未来感十足的包间将以独立灯光和音乐控制系统，随个人喜好调整佐餐氛围，以便让宾客在充满时代感的环境中享受精致而健康的菜品。它还尝试了店中店的模式，在餐饮店中开设一家时尚生活精品店，主售SUBU专属音乐唱片和SUBU风格的餐具饰品，以及来自世界第一手的时尚杂志和家居生活设计品。不同于俏江南其他店的商务气质与Lan Club的奢华风格，SUBU整体设计风格更强调未来感，它更像是一个俏江南的概念馆。

Financial Street Shopping Center restaurant is located in a four-story shop is divided into two parts, the open area under the mall heaven, and right side of the closed zone. Restaurant is taking the international and the young fashion line, cutting-edge of the external design, bold color scheme, so that diners feel like being inside the spaceship.

Throughout the restaurant uses a lot of glass materials, it will be semi-closed state linked to indoor and outdoor. But the glass of the traction of the visual, after all, is the emotional connection, in order to further increase internal and external space with a more rational through the person's emotions and activities of the integration into the designers associated with the bridge, thus derive the arcade of idea - on the basis of the glass and build the structure of the Arcade really created such as the general external capsule form.

Varied colors inspired dining space, unlimited energy, the perfect facial capture the brilliant color of unscrupulous people's sight, lit the entire internal environment. SUBU not just behave in restaurants, but also a new lifestyle concept, futuristic rooms full of light and music will be a separate control system, with the adjustment of personal preference adorned with an atmosphere so that guests in an environment full of sense of the times to enjoy the exquisite and healthy dishes. It has also tried a store-store model, in restaurants, to open a lifestyle store boutiques, the main selling SUBU exclusive music and SUBU style tableware accessories, as well as first-hand from the world of fashion magazines and home design products of life. Unlike other South Beauty restaurant business temperament and Lan Club's luxurious style, SUBU more emphasis on the future of the overall design style, sense, it is more a concept of South Beauty Gallery.

↑半开放式未来主义COCOON风格的餐厅包厢 / Restaurant boxes semi-open style of futuristic COCOON.

↑开敞式用餐区处于商场公共区 / Open-style dining area in the mall public areas.
↓半开放式未来主义COCOON风格的餐厅包厢 / Restaurant boxes semi-open style of futuristic COCOON.

↑SUBU整体设计风格更强调未来感 / SUBU in the overall design style emphasizes a sense of the future.
↓SUBU有着浓郁的梦幻色彩 / SUBU a rich fantasy colors.

A LIN DING MAN XIANG

阿林鼎满香餐厅

06

【坐落地点】西安市碑林区南关正街长安国际广场
【面积】1 200 m^2
【设计】曹成；【设计公司】深圳市汇博环境设计有限公司
【摄影】WISON（泰国）

与以往同类餐厅不同的是，本案业主不要求热火朝天的就餐场面，反而强调安静私密的高端用餐环境。在业主看来，火锅比起同类的高档粤菜，从气氛上更可以拉近宾客之间的沟通距离，并希望设计立意能够强调"冷暖结合，干净素然"的感觉，所以我们自然想到了以冰水与暖色作为切入口。

餐厅主入口为新加建的钢结构廊桥，门廊雨棚由冰凌状的水晶玻璃构成，白色钢构件在超白玻璃间撑起不规则的三角形。由于门廊凹入隐蔽在两栋办公塔楼之间，很不起眼，所以我们利用形态各异的一群流体不锈镜钢企鹅雕塑，将着些许好奇感的客人引上廊桥。步入廊桥雨棚，一眼就看到了雕刻着两朵荷花的水晶玻璃墙面，将最难处理的厨房部分完全隐藏起来。大厅除了吧台外只设置了四围椭圆形的餐台，酒吧台和红酒柜依然沿用了水晶玻璃，就连入口处的两根柱子，也被特制的仿冰块亚克力环抱，浸透着冰色的白光。

为了减弱天花的压抑感，结合主题立意，我们利用镜面不锈钢开模特制了金属水纹板，很好地解决了标高限制这一硬伤。为了增强空间的生动性，

The same restaurant the previous The difference is that this case does not require the owners of the dining scene in full swing, but stressed that the high end of a quiet intimate dining environment. It seems the owners, pot compared to similar high-end Cantonese cuisine, from the atmosphere, the guests can even closer communication between the distance and hoped that the design conception to stress "combination of warm and cold, clean and prime natural" feeling, so we naturally thought of in order to ice water and warm as the entry point.

Restaurant Main Entrance for the steel suspension bridge, canopy composition from the crystal glass, white steel prop up in the glass between the irregular triangle. As the porch hidden indentation between the two office towers, very much, so we use a group of different forms of fluid mirror stainless steel penguin sculpture, will be a bit of a sense of curiosity lead the covered bridges guests. Into the canopy covered bridges, one saw the two lotus flowers carved crystal glass walls, will be the most difficult part of the kitchen completely hidden. Apart from just outside the lobby bar set up all around the oval dining tables, bar sets and red wine cooler is still followed the crystal glass, and even the entrance of the two columns, has also been specially designed acrylic imitation ice surrounded by ice-saturated with color white.

In order to weaken the ceiling of the repression, combined with the theme, we use mirror made of stainless steel metal plates, a good solution to the height restrictions.

↑就餐大厅入口 / Dining Hall Entrance.

↑餐厅入口雕塑 / Restaurant entrance sculpture.

↑入口钢结构廊桥 / At the entrance of the steel covered bridges；↑入口门廊 / Entrance porch.

↑大厅散座区 / Lobby area.
↓平面图 / Plan

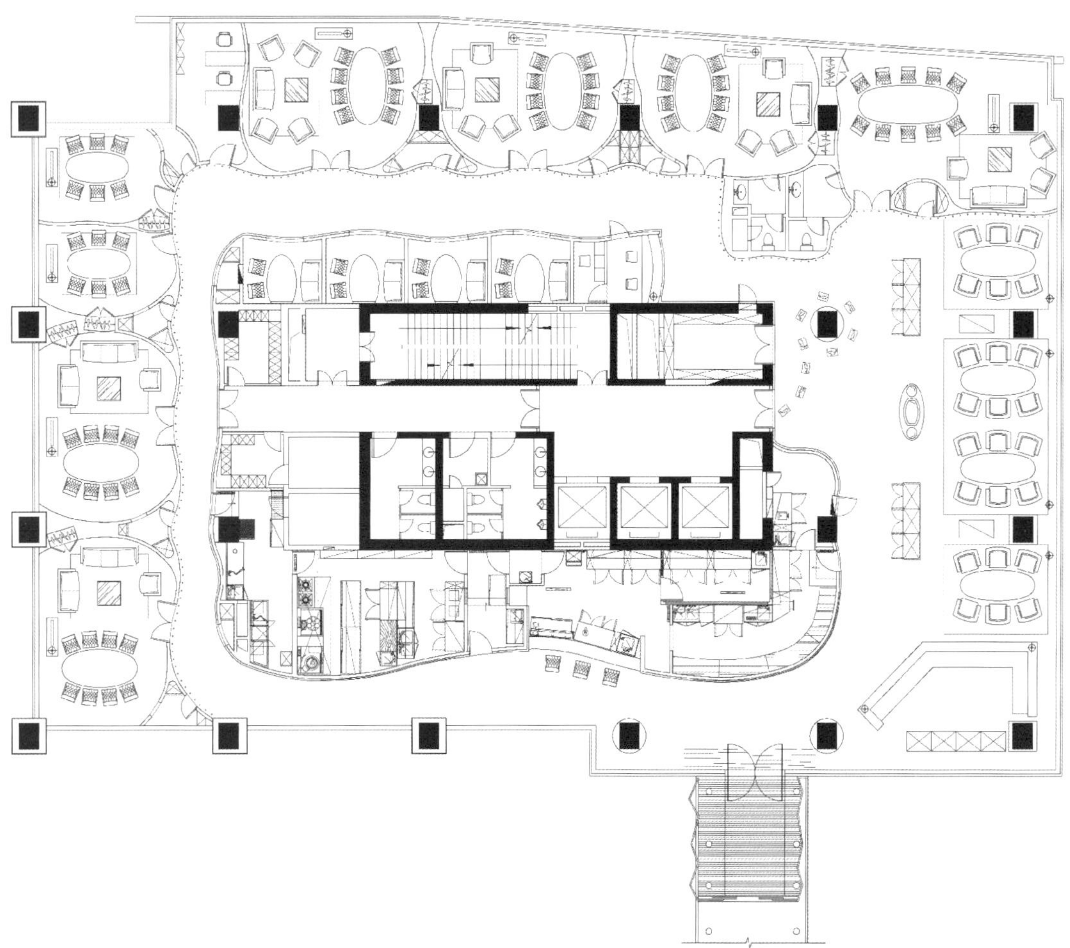

VIP

↑包房入口细部 / Private dining room entrance detail.
←↓不同风格的VIP包厢 / Different styles of VIP box.

SOUTH BEAUTY AT YU XIANG OULU SQUARE

俏江南裕翔欧陆广场店

07

【坐落地点】北京市顺义区天竺镇裕翔路99号；【面积】2 000 m²

【设计】田军；【参与设计】韩玉权、林雨

【设计公司】北京瑞普&田军工作室

【摄影】贾方

本店沿袭了俏江南时尚创新的设计特色，店内饰品大多以木质、瓷器、丝绸等为材料，大量采用隔窗、瓷珠帘等极具中国特色的元素进行装饰，并赋予现代感的创新设计，构成了整个店面的独特氛围。设计师认为，如果仅仅局限于西方的设计语言，中国室内设计师很难在短期内超越西方，只有找到属于中国的设计语言，中国的室内设计才能摆脱传统，走出自己的路。

整个设计选择暖灰色作为主色调，仅仅用明暗对比来区分层次，用色整体而雅致。设计在暖灰色的基调上点缀了中国蓝，其颜色来自中国特有的青花瓷，同样的清新脱俗，简约大方，营造出安静而恬淡的感觉。

餐厅给人的整体感觉是简约而现代的，但是仔细观察不难发现，其中包含了很多中国元素。餐厅以独具江南园林韵味的小桥流水作为入口，灰黄色调中一株盛开的桃花点缀其间，令人眼前一亮。餐厅中以中国传统的兰花图案为装饰母题，这一图案反复出现在椅子、雕塑马、装饰瓷盘、墙面等处；用菱形和方形木格叠加的隔窗分割空间，让人联想到中国的花窗；大厅中的瓷马和其后的座位则象征了中国古代的铜马车。但是无论是青花瓷、官帽椅、小桥，还是陈设物品，都在中国古典形式的基础上进行了抽象和简化，赋予它们以现代感，并且具有了一种含蓄的张力。

South Beauty restaurant continues the stylish and innovative design features, most of the jewelry shop is a wood, porcelain, silk and as material, a lot of use of cut off, bead curtain and other highly decorated with Chinese characteristics, elements, and given modern innovative design And together constitute a unique atmosphere of the entire store. Designers believe that if only in the West, design language, the Chinese interior designer it is difficult in the short term than the West, only to find the design of languages belong to China, China's interior design in order to go beyond the traditional, out of an own way.

Whole design choices warm gray as the main colors used to distinguish between levels of chiaroscuro, the overall color elegant. Design in warm gray tone on the increase in Chinese blue, its color from the special Chinese porcelain, the same delightful, simple and generous, creating a quiet and tranquil feeling.

Restaurant gives the impression that is simple and modern, but look closely you can see which contains a lot of Chinese elements. The restaurant's entrance is the Jiangnan garden style of small bridges, gray peach dotted in the middle, very special. Restaurant with traditional Chinese pattern of decorative orchid theme, this pattern repeated in a chair, horse sculpture, decorative porcelain, wall and other places; with diamond and square wood lattice superimposed split partition space, reminiscent of a flower in China windows; hall of porcelain horse symbolizes the ancient Chinese bronze carriage. But whether it is porcelain, official hat chairs, bridges, or display items, are in China, on the basis of the classical form of abstract and simplified, giving them to modern, and has a low-key tension.

↑用餐处用木框架划分出空间，让人联想到马车 / Dining at wood frame carved out space, reminiscent of carriage.

↑入口走廊 / The entrance corridor.
↓暖灰色木材呈现出古朴的一面，又不失简洁 / Gray wood showing primitive simplicity and conciseness.

↑瓷马及其后面的座位象征了中国古代的铜马车 / Porcelain horse symbolizes the seats behind the ancient Chinese bronze carriage.
↓中国蓝的座椅十分雅致 / China Blue's seat is very elegant.

↑中国蓝和兰花图案作为装饰母题 / Chinese blue and orchid pattern multiple times.
→包厢入口 / Box entrance.
↓半开敞的就餐区 / Semi-open dining area.

俏江南

↑抽象和简化了的中国传统家具中透出现代中国的韵味 / Abstract and simplified Chinese traditional furniture disclosed in the flavor of modern China.
←包厢内陈列的线装古籍，诱发人们更深层的思考 / In the box displayed at wire-bound ancient books to inspire people to deeper thinking.
↓洗手间 / Toilet

XIAO NANGUO RESTAURANT

小南国餐厅

08

【坐落地点】上海新天地；【面积】313 m²
【设计】何宗宪；【参与设计】冯德俊、梁淑婷；【设计公司】Joey Ho Design Limited
【主要材料】横纹玻璃、白色金属框架、白云石
【摄影】柴之澄、鲍世望

本案属于典型海派风格的上海近代建筑。踏进餐厅，不难发现设计师将现代感的时尚与浪漫气息，完美地融进了这座老式建筑。

从门口开始，绚丽的荷花一路从屋顶绽放到地面，经过一个转角，餐厅大堂却又是另一番流动的韵律，屋顶一条窄长雕花玻璃，透射进午后的阳光。设计师以“水”为主题，大大小小水滴形的金属球，借由阳光落到墙面、落到玻璃桌面上，幻化为层层涟漪。“仁者乐山，智者乐水”，中华民族原本视“水”为智慧之源。也只有智者方能在纷繁复杂之间，做出最明智的选择，或者这就是设计师的初衷。

跟一般中式酒楼不同，小南国餐厅没有将焦点放在贵宾包房，反而仿效西方不拘一格的餐厅布局，以一个整体的大空间营造小区间的气氛，着重以穿透的量体及有趣的层次带领客人进入一个宽敞的空间。为丰富餐厅空间，设计师在室内加上各种融会上海古建筑与大自然的特色设计，例如在素花的墙壁和天花雕刻荷花图案，陪衬印有中国国画图案的墙纸，形成有趣和鲜明的对比。此外，设计师又于室内加入亭台的概念，营造柔和与和谐的光影效果。各种巧妙的设计元素结合起来，构成一个明亮、通透的室内环境，将传统上海菜以全新的面貌演绎，为顾客带来崭新与时尚的餐饮体验。

This case is a typical Shanghai Style, Shanghai's modern architecture. Enter the restaurant, not difficult to find designers and contemporary fashion and romantic atmosphere, perfect to melt into this old building.

Into the gate, a beautiful lotus flowers blooming all the way from the roof to the ground. After a corner, dining hall and yet another scene in the afternoon sun radiation coming from the roof. Designer of "water" as its theme, large and small water droplets to form metal balls, the glass fell on the desktop, issued layers of waves. "To be benevolent because the mountain become happy, wise man happy water", the Chinese had considered the "water" as the source of wisdom. Only the wise can be very complex and complicated world, and make the most informed choice, and this is the designer's original intention.

And the general Chinese restaurant is different from the case did not focus on the VIP room, but the layout of the restaurant, like the west, to a large space to create a whole district between the atmosphere, with emphasis on the quantity of penetrating body and the level of fun to lead a guest into a spacious space. In order to enrich restaurant space, interior designers with a variety of findings derived from the ancient architecture and nature Shanghai features the design, for example, prime the walls and ceilings carved flower lotus pattern, printed with Chinese painting wallpaper patterns to create interesting and in contrast to . In addition, designers to join in the indoor pavilion concept, to create a soft and harmonious lighting effect. A variety of clever design elements combine to form a bright, transparent indoor environment, the traditional Shanghai cuisine is a new face displayed for customers to bring new and stylish dining experience.

↑大大小小的水滴形状的金属球在空间中散落 / The shape of water droplets, large and small metal balls scattered in space.

↑绚丽的荷花一路从屋顶绽放到地面 / A beautiful lotus blossom from the roof to the ground.
↓设计师以“水”为主题 / Designer of "water" as the main element.

↑散座区 / Hall area.
↑天花顶上的荷花烘托主题 / Ceiling on top of the lotus contrast environment.
↓平面图 / Plan

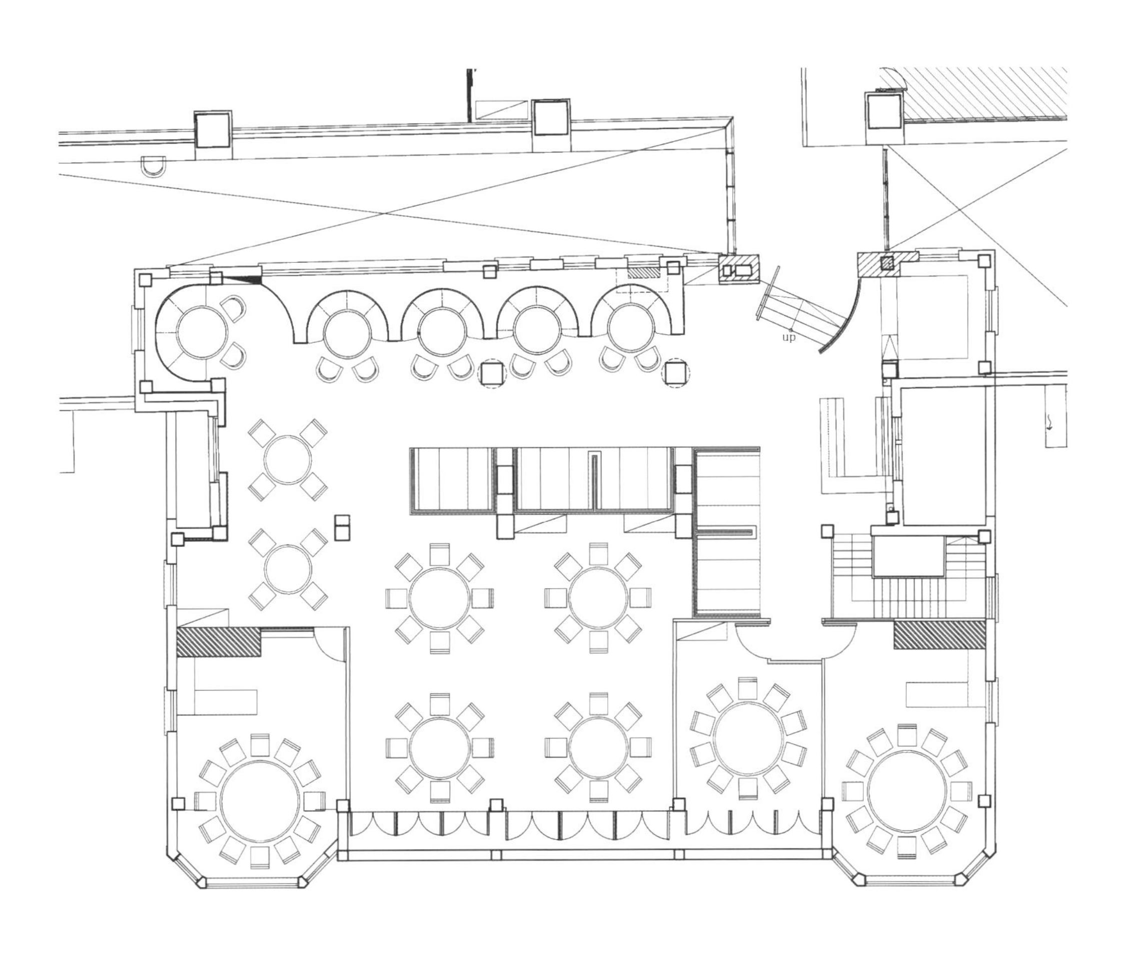

↑厅内红色、白色的流苏仿佛一道水幕 / Hall red and white curtains like a water curtain.
←白色的流苏圈出老上海的浪漫气息 / A white curtain circle the romantic atmosphere of old Shanghai.
↓包厢的风格与大厅一脉相承 / The same style of box and halls.

HONEYCOMB RESTAURANT

深圳蜂巢概念餐厅

09

【坐落地点】深圳市南山区（深圳保利剧院内）；【面积】1 300 m^2

【设计】迫庆一郎〔日〕；【设计公司】SAKO建筑设计工社

【主要材料】亚克力板、FRP (Fiber Reinforced Plastics)

【摄影】松美佐江

所谓餐厅，就是把生活中最基本的饮食行为进行外部化的场所。本案追求的是把空间处理成——非日常空间。把"蜂巢"作为"非日常空间"的名字，亦有让客人在享受食物的同时也"享受空间"的意思。

本案外壁的幕墙是半镜面的，夜幕降临时，镜面倒映着白色面，使内部空间呈现出犹如被波浪环绕着的梦幻一般的景象。由外面看来，光线从白色大曲面的无数个"洞"中透出，使人仿佛看到一个巨大生命体般的感觉。白色面内侧有两层，由大厅和包厢两种空间类型组成。入口处纵贯了"小空间"，从那里穿过就是"大空间"了。与许多中式餐厅一样，这里既有就餐大厅，也有包厢。如何塑造出这两个不同性质的空间是设计中最重要的环节。我们在"大空间"用白色不透明的面，"小空间"用亚克力透明的面，作为界面来分隔各个空间。

"小空间"的透明面，逐渐变形成波浪状，上面开了许多洞。其外侧是调暗了光线的入口空间，内侧是大厅。黑色大理石地面、镜面不锈钢顶棚、透明亚克力墙壁等反光材料的组合，模糊了入口空间的轮廓，让其中的客人在享受美食的同时，有眺望空间的感觉。

"大空间"的白色面也一样，在平面上呈波浪起伏状，生成5条"褶皱"状的分界线。从立面上看，这5条"褶皱"分割出6个凸起的空间，它们像"巢"一样存在着。

Restaurant is the most basic diet of people living outside of the conduct of a place. Pursue the case is to spatial processing into a - non-routine space. The "Honeycomb" as a "non-everyday space," the name also allow guests to enjoy food, but also "enjoy space" means.

This case the outer wall of the curtain wall is a half-mirror, and night, mirror reflected upside-white face, so that interior space showing the waves like a dream around the general picture. It seems from the outside, light from the white surface of the numerous large "hole" in revealing that people seem to see a huge body of life-like feel. Inside the two-story white face, from the halls and private rooms are two room types. At the entrance to north-south "small space", and from there through is the "large space" of. Like many Chinese restaurants, where the existing dining hall, there are boxes. How to mold the nature of these two different design space is the most important aspects. We are in the "large space" using a white opaque surface, "small space" using a transparent acrylic surface, as the interface to separate the various spaces.

"Small space" transparent surface, gradually changing the formation of wave-like, the above has many holes. Lateral is the entrance hall on the inside. Black marble floor, mirror stainless steel roof, transparent acrylic walls, a combination of reflective material such as blurred outline of the entrance space, so that one of the guests enjoying the food at the same time, it overlooks the feeling of space.

"Large space" white face, in the plane wave-like ups and downs to generate 5 "fold" shape of the line. From the elevation point of view, this 5 "fold" Split out of six convex space, they are like a "nest" the same.

↑VIP室，白色面构成的“巢”，其直径、倾斜度都各有不同 / VIP room, a white face constitutes a "nest" and its diameter, tilt are different.

↑几个“巢” / Several "nest."

↓白色面构成的“巢”，其直径、倾斜度都各有不同 / A white face constitutes a "nest" and its diameter, tilt are different.

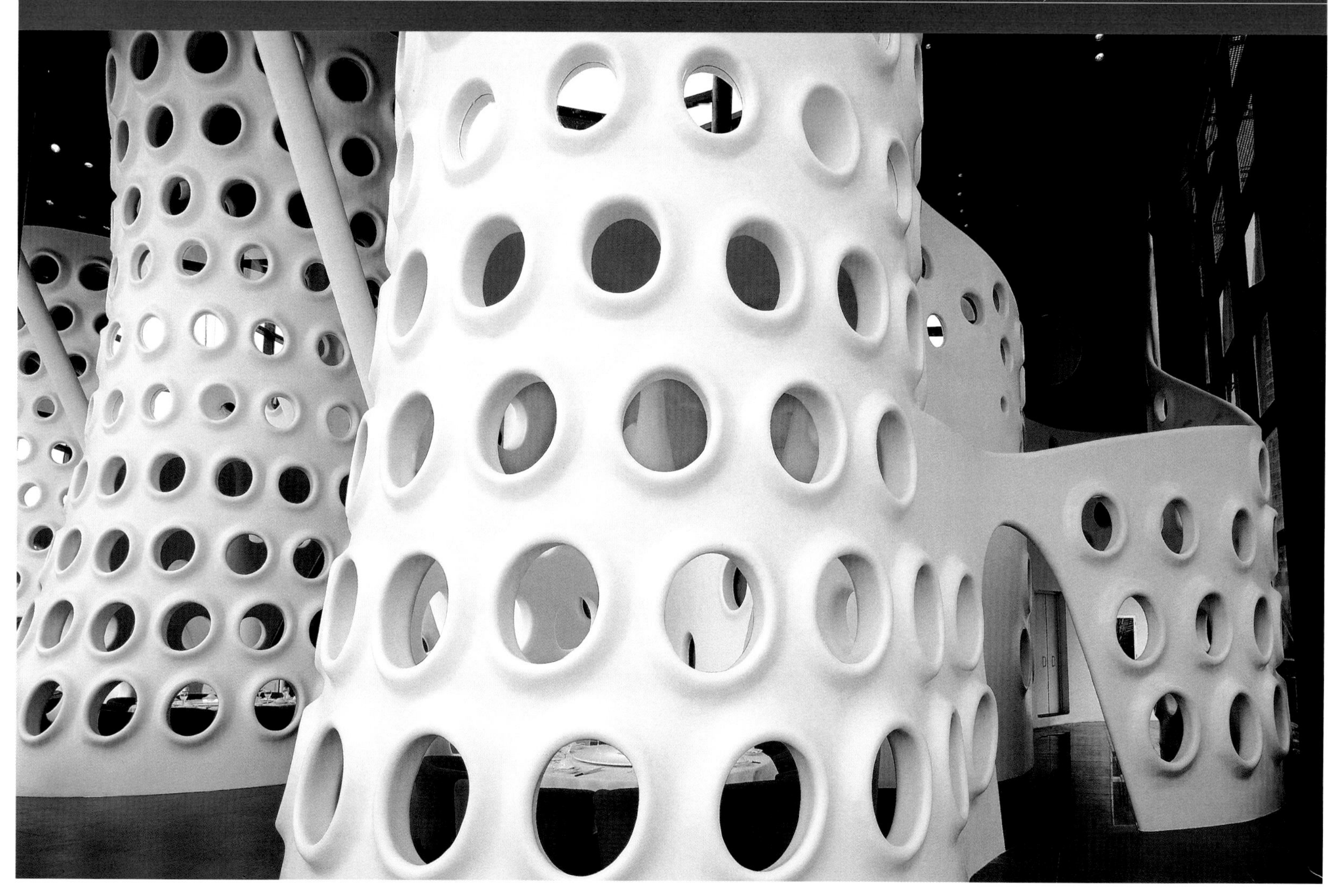

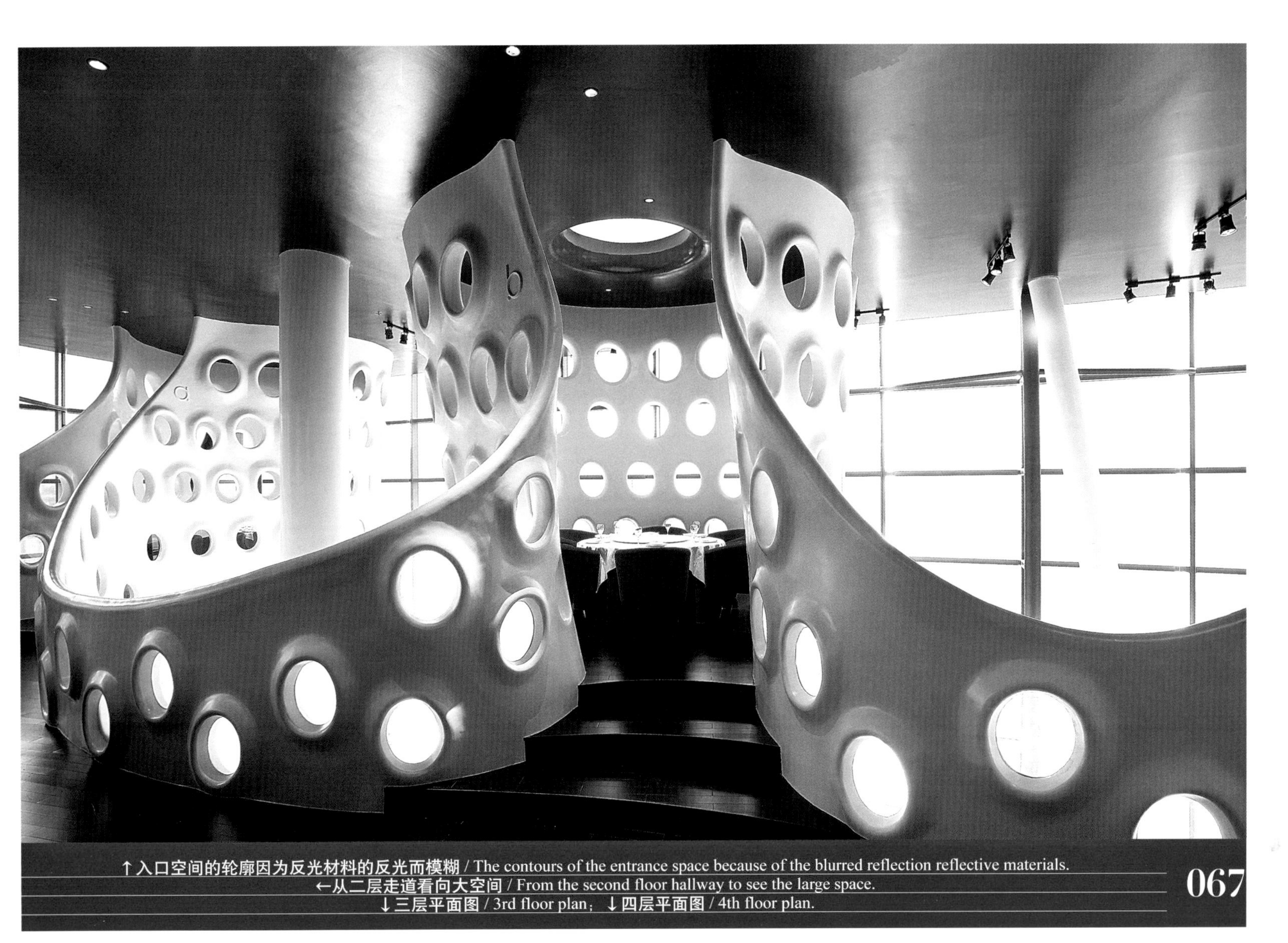

↑入口空间的轮廓因为反光材料的反光而模糊 / The contours of the entrance space because of the blurred reflection reflective materials.
←从二层走道看向大空间 / From the second floor hallway to see the large space.
↓三层平面图 / 3rd floor plan; ↓四层平面图 / 4th floor plan.

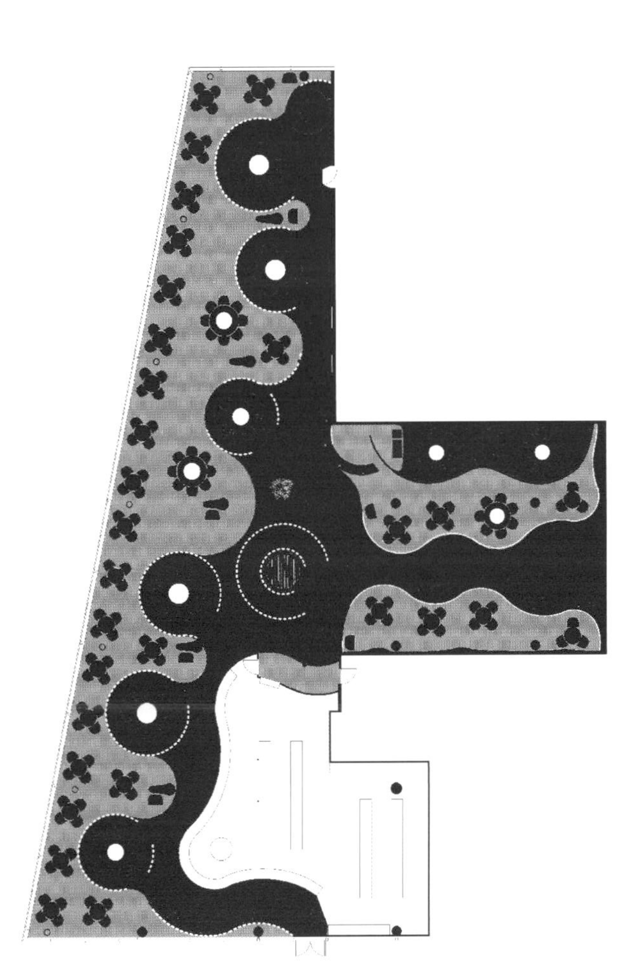

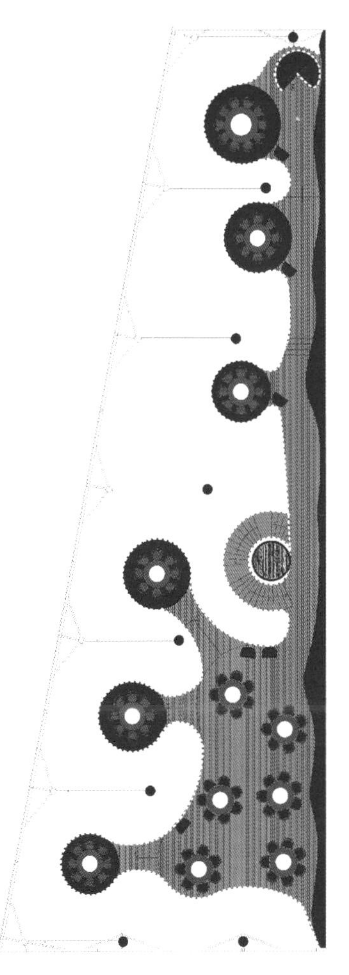

SI CHUAN CUISINE RESTAURANT

上海麻辣诱惑餐厅

10

【坐落地点】上海长宁区长宁路龙之梦购物中心7楼7-211；【面积】1 240 m^2
【设计】耿治国；【设计公司】飞形设计事业有限公司
【主要材料】黑红色条纹绒布、红白色夹膜玻璃、灰色纱帘、黑色明镜、黑色板岩砖、银色皮革
【摄影】Charlie Xia

整个餐厅为全开放空间，包括入口、大堂、开放性吧台及用餐区，依靠设置隔断和控制家具尺寸保证餐位的私密性和距离。设计师通过舞台式就餐区（Horizontal Stage Dining）很好地避免了大空间的呆板无趣，在大空间里用台阶高低来实现不同就餐区域的划定，增强领域感的同时，令空间充满趣味性。

在空间的灯光布置上为：点、线、面光源依餐桌的布置而游走，突破常规，功能性和私密性统一；在非就餐区适度调低照明亮度，满足人对私密性的心理需求。斑驳于柱间墙壁的光影和浓艳的红黄绿紫的大色块，毫无疑问地说明设计师是光色高手。不锈钢、透光亚克力、珠帘等材料的使用，拉近了餐厅与年轻消费群体的心理距离。

设计师推崇跨界整合设计，透过发光地台上餐椅靠背流畅的线条、从地面向上生长的隔断、卷曲的造型就像是在欣赏合奏的麻辣交响乐谱，多变而和谐，婉转而流畅，强调空间的独特个性和家具的趣味性。跨界整合设计使多种类艺术相互影响和渗透，艺术灵感与商业策略深度融合，激发了更新、更有挑战性的创意和更高境界的设计水平。

The entire restaurant for the whole open space, including the entrance, lobby, open bar and dining area, relying on set furniture, partition and control the size of dinner place to ensure privacy and distance. Horizontal stage dining well by designers to avoid a large space, rigid boring, in large space with high and low level to achieve the delineation of the different dining areas and enhance a sense of the field at the same time, so space is full of fun.

Lighting arrangement in space for the: point, line, surface light source in accordance with the table layout of the walk, breaking routine, functionality and privacy reunification; a modest reduction in non-dining areas lighting levels to meet the people's psychological needs for privacy . Mottled light and shadow on the wall between columns and a rich ruby red yellow and green and purple large blocks of color, there is no doubt explains designer is light color master. Stainless steel, translucent acrylic, bead curtain and other materials used to narrow the restaurant with the younger consumer groups psychological distance.

Designers respected cross-border integration of design, through the light dining table back to smooth lines, cut off from the ground upward growth, curled shape is like a symphony in their appreciation of spicy ensemble spectrum, varied and harmonious, tactfully and fluency, emphasizing the unique character of space and furniture interesting. Cross-border integration of design allows a wide array of artistic interaction and penetration depth of artistic inspiration and business strategy integration, inspired newer and more challenging ideas and a higher realm of design level.

↑抬高的地坪整体发光变色，塑造出舞台的戏剧效果 / Elevation of ground floor as a whole luminous color, shape out of the stage drama.

↑全开放的空间 / All open space.
↓餐椅布置形式多样 / Chair arrangement in various forms.

↑独特的包厢造型 / A unique box shape.

平面图 / Plan

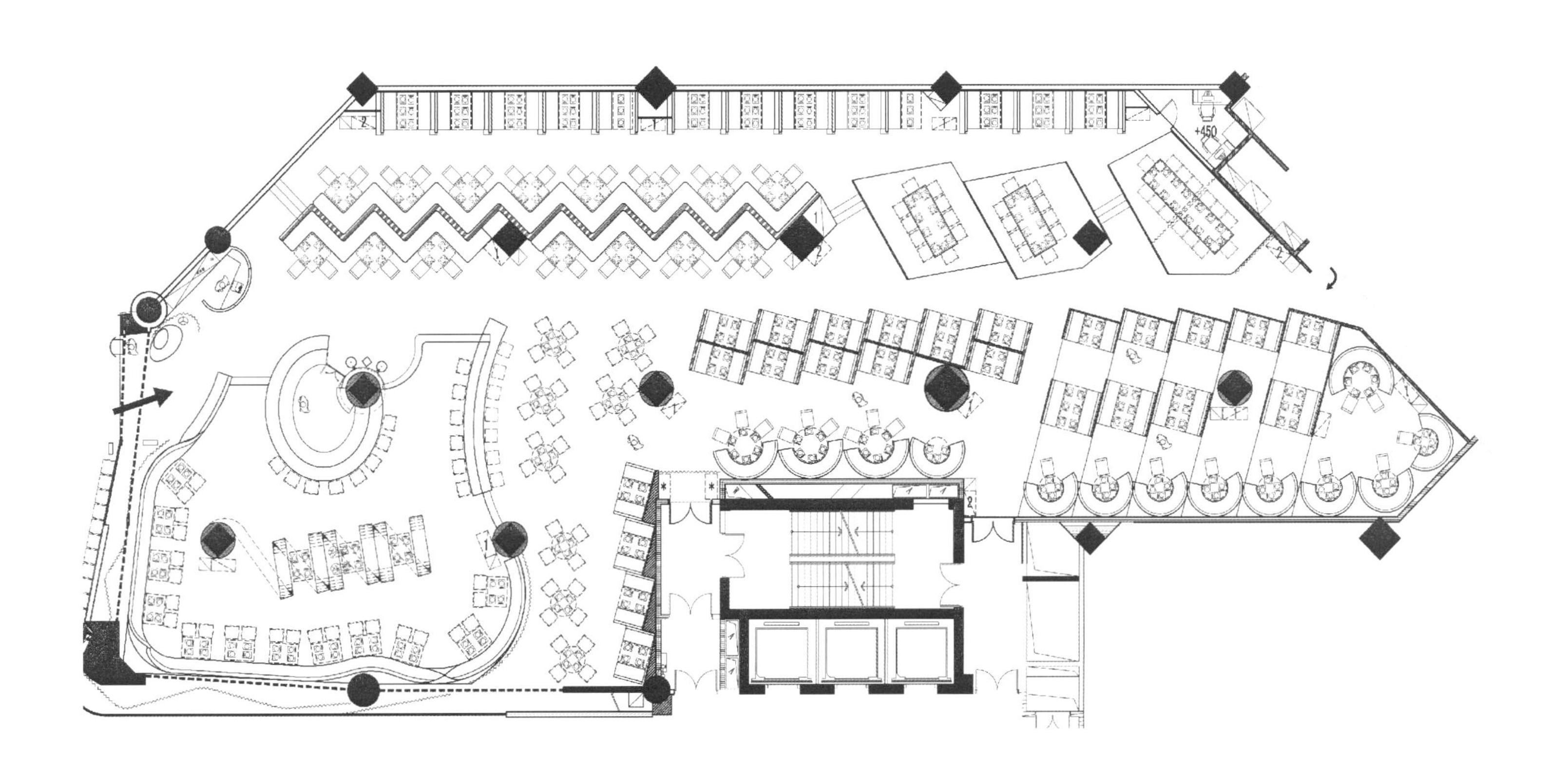

A FAIRYLAND IN HANGZHOU RESTAURANT

外婆家杭州众安店

11

【坐落地点】杭州中山北路；【面积】1 000 m^2

【设计】戴朝盛

【设计公司】杭州山水组合设计公司

【摄影】贾方

在这个大空间里，设计师出于对中国传统文化的深厚情结，运用简练的手法，把中式花窗重新定义。镂空的造型因视线的穿越而赋予变化，同时运用无彩色系的黑色为主调，缀以有光泽的金色，含蓄而有力地表达一份现代中式意境。两者相鉴，这种并置的异质文化在设计师的精心设计下，因为写实、丰满和强调明暗对比，紧密地融合在一起，饱和地向人们展现出餐厅设想的文化主题，从而形成了与众不同的富有个性的建筑室内设计。

从整体色调上来看，本店以深木色和少许的金色为设计主线，作为装饰的母题，传统图案纹样的木雕被大面积的运用，在门、窗、隔断、家具、装饰品上重复出现。而与之相配的黑色皮质坐椅，现代感的金属帘饰和吊灯，精致奢华，简洁婉约，这种强烈的视觉效果让人不禁有种跨越时空之感。

用象征富贵的牡丹图案作雕刻装饰，是空间的点睛之笔，最能体现空间的品味与审美。嵌以玻璃窗上的牡丹群组图与屋顶上金色浮雕的牡丹图都采用了相同的题材，将客人包裹在流光溢彩之中，使该空间的庄重、热情、豪华、雅致无限延伸，奏出一个小高潮，让人不禁油然而生华丽尊贵之感，同时也实现了中西两种艺术形式在实用层面的最佳融合。此外，传统文化特色十足的个性工艺品摆件妙趣横生，为此景平添几分生气。精心布置的鸟笼摆设，点缀桌面，为整个空间带来浓厚的人文气息和怀古幽情。

In this large space, the designers out of the profound Chinese traditional culture complex, using concise techniques to bring Chinese-style flower windows to be redefined. Hollow shape conferred by eye through the changes, while the use of black as the main colors, decorated with shiny gold, subtle and powerful expression of a modern Chinese-style mood. In comparison, this culture in the designer's well-designed, because realism, fullness and emphasis chiaroscuro, closely together, saturated a restaurant to show people the vision of cultural themes, thus forming a distinctive Architectural interior design.

The overall tone point of view, restaurant with dark wood color and a little gold for the design of the main line, as a decorative motif, traditional patterns are large-scale patterns of wood use, in the doors, windows, partitions, furniture, decorations on the repeated . The match with the black leather furniture and modern decorative metal screens and chandeliers, exquisite luxury, simple and graceful, this makes a strong visual impact can not help feeling a kind of across time and space.

With the peony symbolizes wealth and patterns for carving decoration, best embodies the tastes and aesthetic space. Embedded in the window of the Peony Group Photos with gold embossed on the roof of the Peony map have adopted the same theme, will be among the guests wrapped in, so that the space solemn, warm, luxurious, elegant infinitely extended, merged a small climax, people producing beautiful noble sense of the health care, but also realized the two kinds of Chinese and Western art forms in the practical level, the best convergence. In addition, traditional cultural characteristics of the craft ornaments add a bit angry for this scene.

↑极富特色的吊顶设计使包厢韵味十足 / Highly distinctive ceiling design allows the full flavor box.

↑在静谧的空间中，一对鸟笼被置于入口 / In the quiet space, a pair of cages are placed in the entrance.

↓弧形沙发和弧形牡丹刺绣相映成趣 / Peony embroidery curved sofas and arc set off each other.

↑隐约透着包厢里另外的空间 / Vaguely be able to see another layer of space inside the box.
↓平面图 / Plan

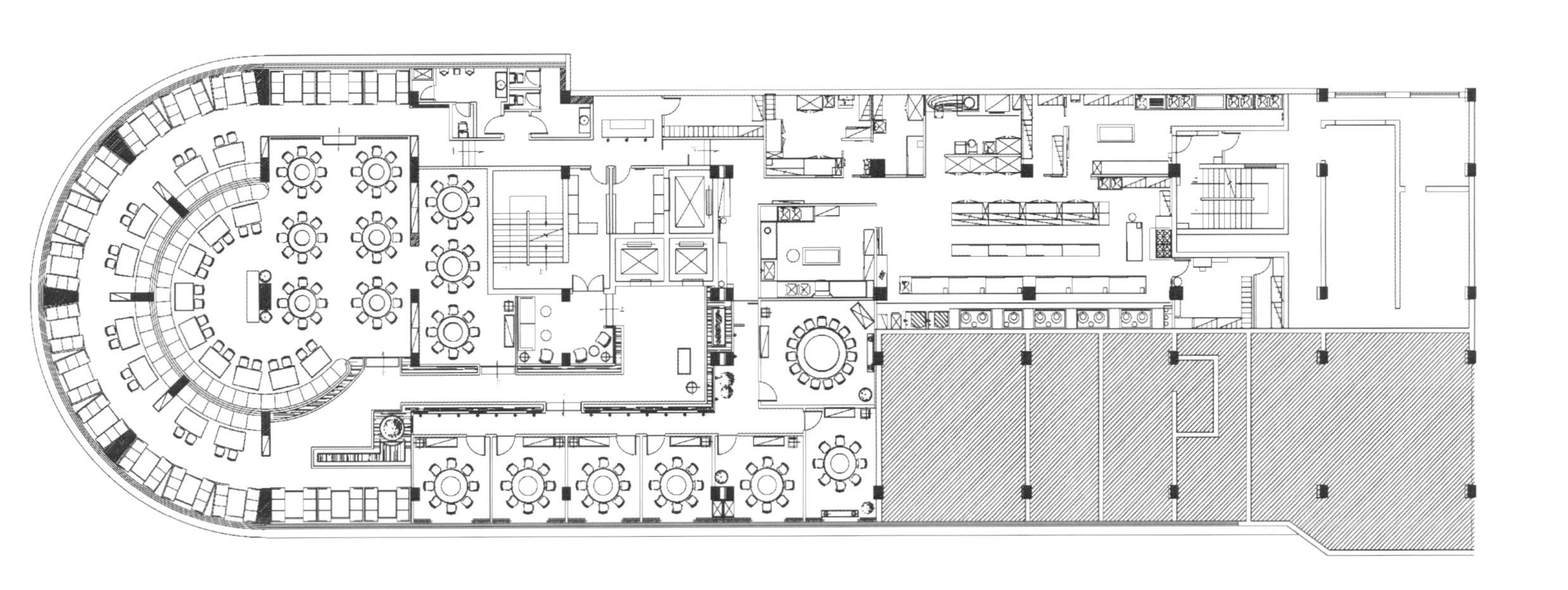

↑硕大如斗拱形状的“绿如意”装饰在入口进门处 / Enter the gate at the entrance, huge arched "green Ruyi" decoration.
↓福鹿名肴会入口 / Fu Lu Restaurant entrance.

↑几件很中国的摆设，红色的桌旗使空间倍增时尚感 / A few pieces of very Chinese decorations, red silk to make room for additional fashion sense.
↓镂空花格以及金属帘设计丰富了走廊的景致 / Pierced grille and metal screens designed to enrich the landscape corridors.

ROUND TOTEM

味腾四海火锅店

12

【坐落地点】云南昆明市；【面积】2 000 m^2
【设计】张渝；【设计公司】昆明思成工程设计有限公司
【主要材料】花岗石、水洗石、玻璃、橡木
【摄影】蒋舒宇

火锅的最大特色就是——沸腾。而"圆"形很好的契合了这种沸腾的构思，同时"圆"所表现出来的无限广阔的包容性，极像昆明人缓慢而自在的生活态度。由此可见，将"圆"视为本店的图腾也就再适合不过了。

这个店原为四层平层，每层500 m^2，平面呈四分之一圆形，一层略高，其余楼层均为梁底2.4 m。原来只有一、二层营业，业主决定把三层也纳入营业范围，这样不仅增加了食用空间，而且充分利用了原有楼房的格局，可谓一举两得。但是它同时带来了一个问题——空间的改造已不可避免。为了保障整体空间的品质，避免空间过度单调压抑，必须要拆掉二、三层的部分楼板。通过一系列特殊的处理方式，空间由此变的大气而生动，环绕包间的环状金属带也越发的加强了升的态势，不仅丰富了整个空间层次，也使得就餐环境更为典雅尊贵。中庭的这两个包间也是客人比较喜欢的房间，借中庭梁的金色与包间绿色玻璃呈现昆明阳光和春城的印象。

圆是这个空间中出现最多的元素，在造型中大量出现的圆是鲜活的，流动的。外立面用上升的气泡组成，室内实木隔屏上也是相同；中庭的环状金属带扭动上升，圆形木帘飘动摇曳。圆形餐桌、圆形餐具、圆形灯具、圆形饰物，"圆"暗示出沸腾火暴的场景。

This change implies a qualitative sublimation of things associated with the development of the scene, with "round" shape to represent the meaning of this increase was originally conceived as a designer.

The shop was originally four flat layers, each layer 500 square meters, was fourth circular plane. A layer of slightly higher at the end of the remaining floor beams are 2.4 m. Previously, only one or two-story business, the owners decided to also include the three-tier business, so that not only increased the consumption of space and full use of the pattern of the original buildings. But it also gives rise to a problem - the need for space transform. In order to protect the quality of the overall space to avoid excessive monotony of space must be removed two or three-storey part of the floor. Through a series of special processing methods, space and lively atmosphere of this change, around the ring between the metal strip packages are also increasingly the trend of the strengthening of the liter, not only enriches the entire space level, also makes an elegant dining environment more honorable. Atrium between the two packages is also the guests prefer a room, through the glass, showing the sun and Spring City in Kunming impression.

" Round " appears in this space is the largest element in modeling the proliferation of round is fresh, and mobile. Facade with the rising air bubbles formed, indoor wood Partition is also the same; in the atrium of the ring with a twist of metal rose, round wood curtain fluttering. Round dining table, round tableware, round lamps, round ornaments, " round " implies a boiling hot scene.

↑一层大厅空间的虚实变幻 / Floor lobby space and the reality changes.

↑一层大厅 / Floor lobby.

↓中庭包间是空间的主角 / Hall of space between the middle of the package of visual center.

↑一层大厅 / Floor lobby；↑波浪形天花 / Wave-shaped ceiling.
↓一层平面图 / Floor plan.

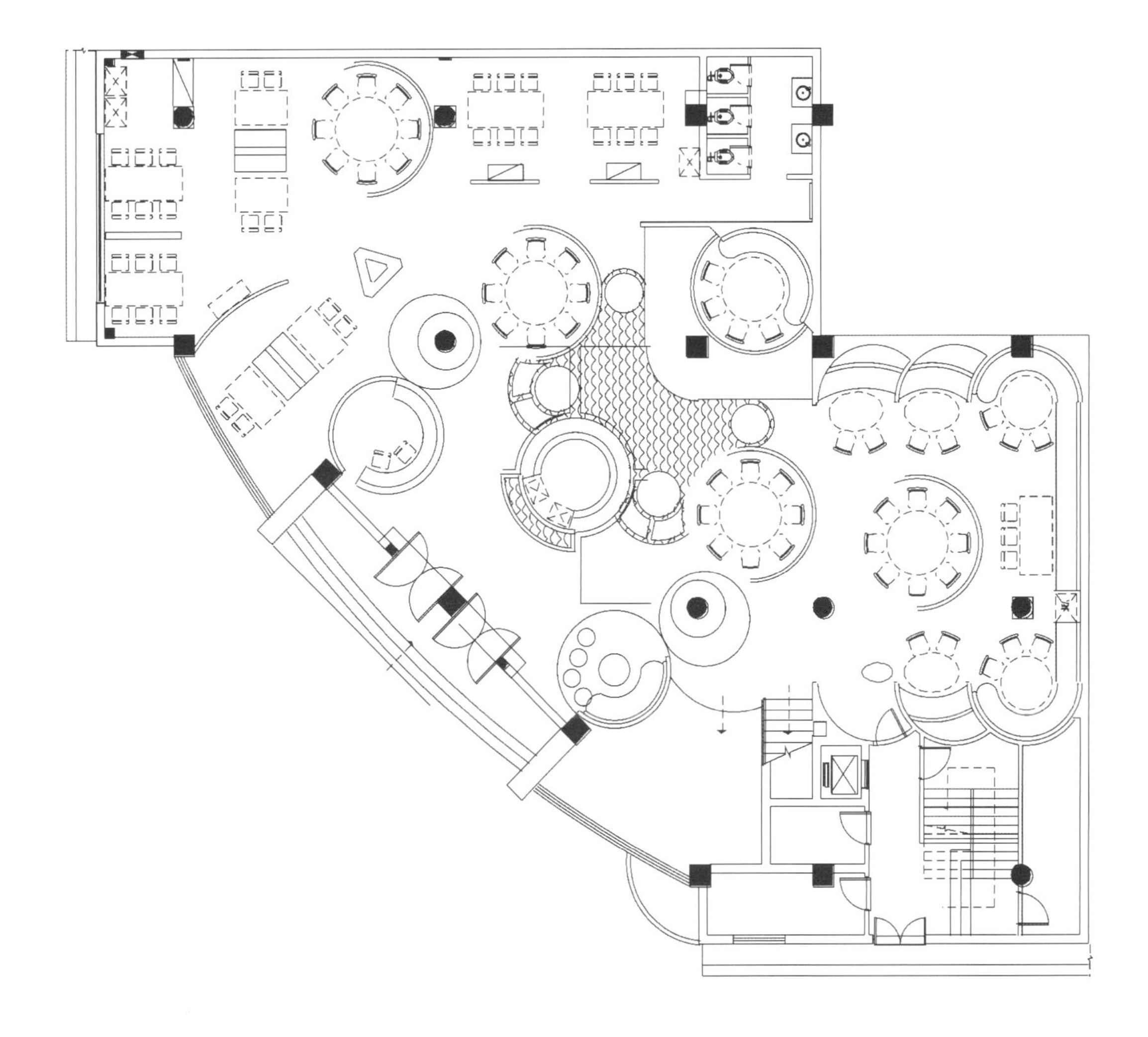

↑包厢 / Box
←中庭包间是空间的主角 / Hall of space between the middle of the package of visual center.
↓二层平面图 / 2nd Floor Plan.

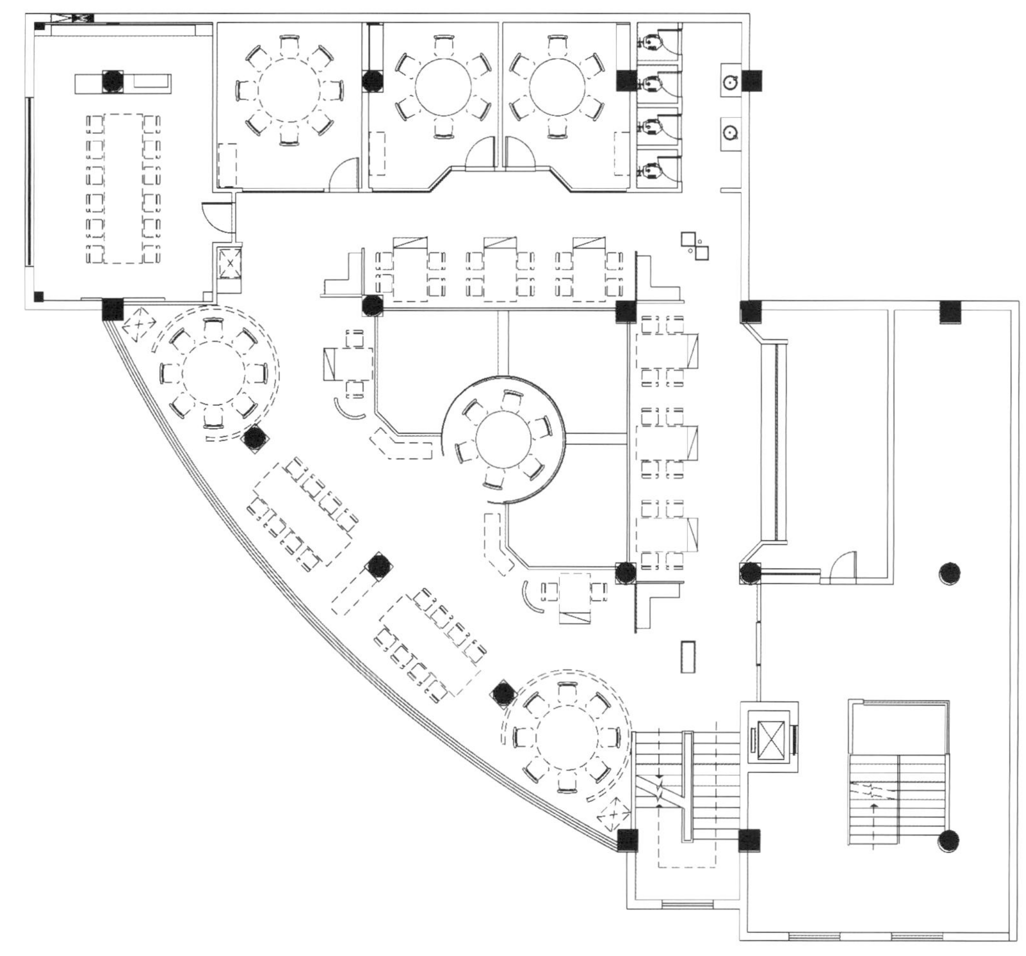

A CHINESE RESTAURANTIN XUZHOU

徐州某餐厅

13

【坐落地点】江苏徐州市；【面积】1 114 m^2

【设计】Horace Pan，Alan Tse，Vivian Chan

【设计公司】泛纳国际设计顾问有限公司

【摄影】Ng Siu Fung

原建筑中并没有将上下两层直接连接起来的楼梯，因此需要设计师在餐厅的内部设计出一个垂直的楼梯，以联通上下两层；另外在于餐厅是闹市区，周围的环境比较嘈杂，楼梯的存在能起到良好的隔音效果。

整个"笼子"由通高的深色木条拼接构建而成，木条之间镶嵌了若干同质同色的小木块，这些小木块呈锯齿状排列，好像是嵌在墙壁里的楼梯扶手，其起伏的形状带给客人一种向上攀登的动力。隐藏在楼梯之间的照明设备为空间带来了一种虚实相间的莫名体验，客人在上楼的过程中，仿佛浮游在一个神话般的世界中，连原本沉重的脚步似乎都变得轻盈起来。

不同色调所蕴含的特殊意义被设计师用来作为划分空间层次的标准。位于一层的酒吧是明黄色，明黄色的发光吧台是空间的视觉焦点；大型的吊灯更为空间增加了一道独特的风景。二层的主色调是暗红，暗红色的圆形半开放包间以黑色的纱幔分隔，条状的天花中隐藏着暗红色的灯光照明，从视觉上营造出一种沉静的就餐氛围。开放式就餐区部分的天花采用了点状暗红色灯光，将下面的白色桌椅、白色隔断晕染出淡淡的红。为了与整体空间色调保持一致，VIP包厢的外墙也采用了暗红色，当然包厢内部则是另外一种全然不同的体验了。

The case of the original building, and did not directly link the two up and down the stairs, requiring designers to design in a restaurant inside a staircase; another is that restaurants are downtown, the surrounding environment is noisy, the existence of the stairs can play good soundproofing.

The whole "cage" by-pass the high dark wood mosaic is built, a number of strips of wood inlaid between the homogeneity of small pieces of wood the same color, these little pieces of wood serrated arrangement seems to be embedded in the wall where the staircase handrails, which kinds of shape, a kind of upward momentum brought to the guests. Hidden in the staircase lighting equipment for the space between an actual situation has brought experience, guests upstairs in the process, as if floating in a fabulous world, even the heavy footsteps of the original seems to have become light up.

Different shades of meaning inherent in the particular division of space has been used as a designer-level standards. The ground floor bar is bright yellow, bright yellow light-emitting space bar is a visual focus; large chandelier is more room for an increase of a unique landscape. The second floor of the main colors are dull red, dark red circle with a black gauze rooms separated by strips of smallpox in the hidden dark red lighting, from the visual to create a serene dining atmosphere. Part of the open dining area with point-like dark red ceiling lights, following the white tables and chairs, white partitions reflecting a touch of red. In order to be consistent with the overall tone space, VIP box also used the external walls of dark red, of course, is another box inside a different experience.

↑楼梯踏步中隐藏的灯光使楼梯间变得轻盈灵动 / Stair stepping hidden lighting to light the stairwell to become smart.

↑巨型吊灯由数百只白色的飞鸟组成 / Chandelier from the composition of hundreds of white birds.
↓楼梯间像一个飘浮在空中的笼子 / Staircase floating in the air like a cage.

↑从楼梯间出来进入开放式就餐区 / From the stairwell out into the open dining area.
↓二层平面图 / 2nd Floor Plan.

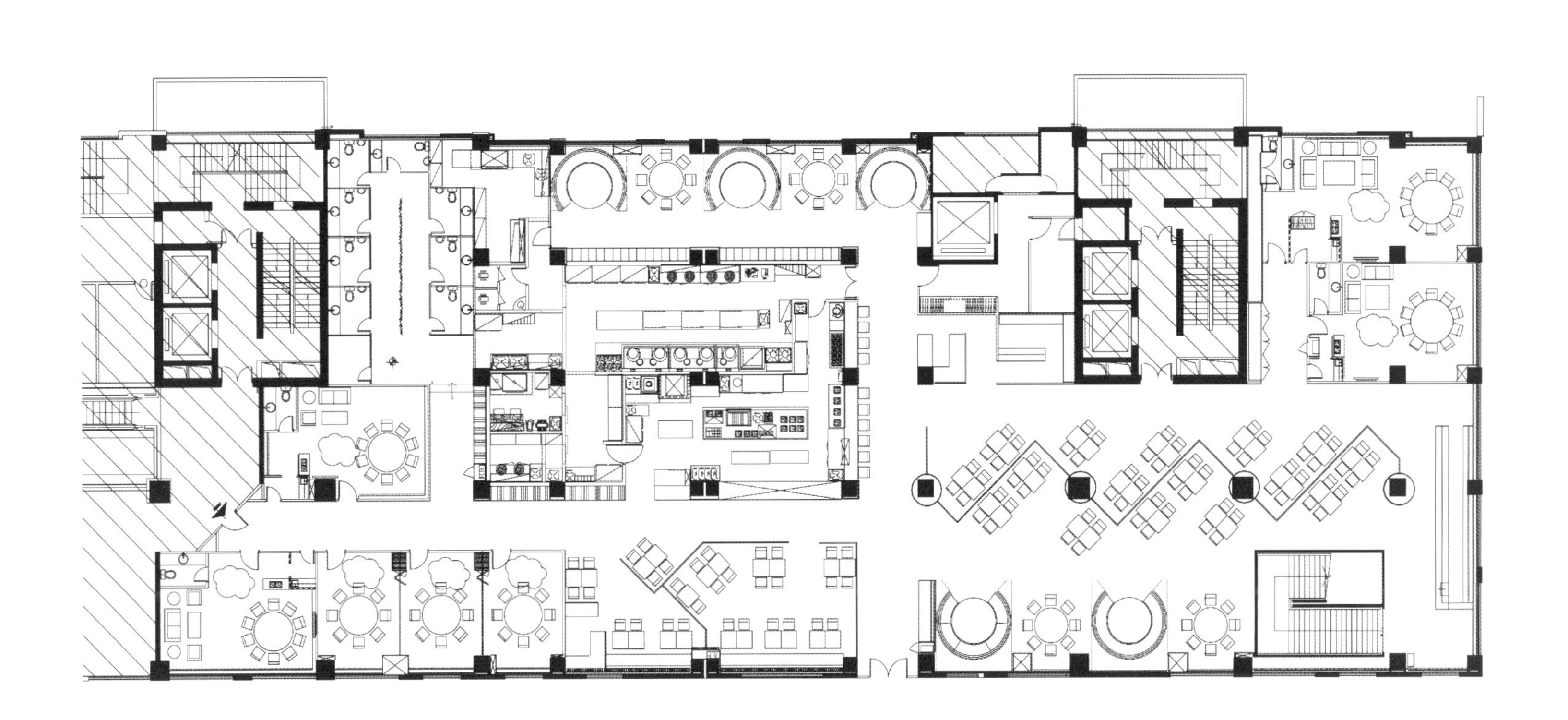

↑在就餐区的隔断上嵌入很多白色金属方块 / The walls in the dining area to embed a lot of white metal box.
↓暗红色开放式包厢 / Dark red open box.

↑VIP包厢内景 / VIP box interior.
↓山水画般的洗手间 / Chinese paintings on the bathroom wall.

YONG YI CHINESE

永颐中餐厅

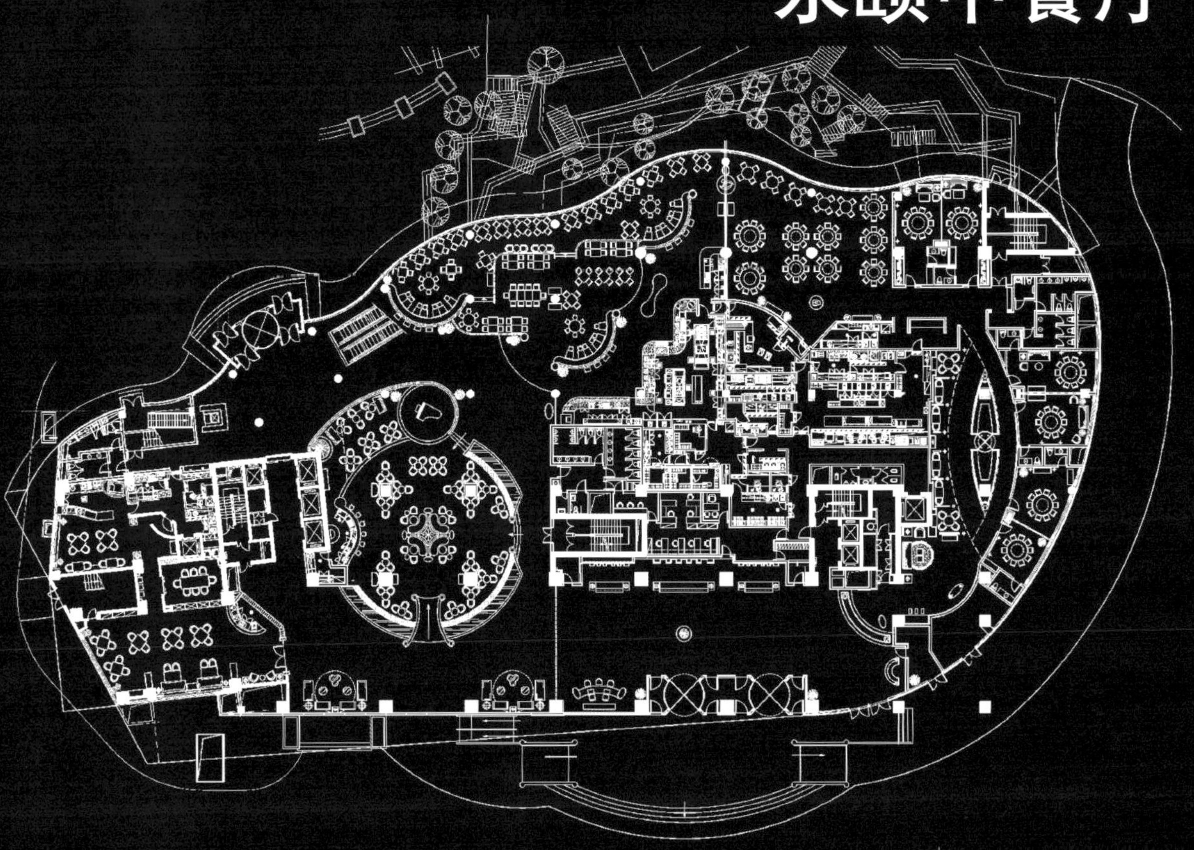

14

【坐落地点】北京海淀区远大路25号
【面积】561 m^2
【设计】香港KWSG
【摄影】贾方

本案是酒店里一所提供中西方特色美食的餐厅，设计师有意将路线设计得迂回曲折，借鉴了中国园林曲径通幽意境的同时，又通过绚丽色彩的灯光营造出西方浪漫的夜色风情。吊装的钢结构天花和斑斓的顶灯设计，大面积的落地玻璃窗和中式木框隔断的造型结合光影效果，让天花有了浮动的不真实感，宛如安谧静异的星空。

大厅中，扇形吊顶下的灯饰处理也别具风格，灯影重叠交错，明灭绚丽，穿插的植物纹样装饰不经意间散发着中国式的神秘。餐厅中青花瓷图案的古典坐椅安静地围置在绛红色的圆桌旁，白瓷盘中葱茏的水仙似有还无地散发着芳香。

"静、洁、雅"是永颐中餐厅空间内涵表达的重点。中国古典美学强调意境的创造，意境是一种情景交融、虚实相生的形象所诱发和开拓的审美想像空间。意境的构成是以空间镜像为基础的，是通过对镜像的把握达到"情与景汇，意与象通"。全透明的落地玻璃幕墙将室内外隔开；老式深色餐桌，总会让人想起过去的岁月；青花瓷图案的椅子打破了空间的沉闷；花色图案简洁的地毯为空间增添了一份精致和奢华。

This case provides an hotel features Chinese and Western cuisine restaurants, designer intends to line twists and turns designed to learn from the Chinese conception of the garden as well as through a beautiful romantic lighting create a Western style. Steel ceiling and gorgeous dome light design, large areas of floor windows and cut off Chinese-style wooden frame shape out of light and shadow effects, making like a quiet and starry ceiling.

In the hall, the fan-shaped ceiling lights under the handle is quite distinctive, colorful lights, decorated distribution patterns of plants interspersed with Chinese-style mystery to it. Restaurant chair with a porcelain pattern quietly Circuit in the round table next to the porcelain of the narcissus aroma exudes.

"Quiet, clean, elegant" is the dining room to express connotation of the point. Chinese classical aesthetics emphasized the creation of mood, mood is a kind of scenario integration, actual situation and explore the intersection of aesthetic space. The composition of conception is based on space-based, to achieve "integration of intelligence with the King, Italy, and as the similarities." A transparent glass wall separating the indoor and outdoor; old dark tables, reminiscent of the past years; porcelain chair to break the pattern of space dull; simple carpet for space adds a refined and luxurious.

↑餐厅路线的设计避免了因面积不大而带来的拘谨和狭窄 / Restaurant routes designed to avoid small area brought about by the narrow.
←一层平面图 / Floor plan.

↑餐厅借鉴了中国园林曲径通幽的概念来渲染意境 / Restaurant draws on Chinese gardens in mood.
←中国园林中的借景手法自然运用在座位之间的木隔断上 / Chinese garden design in the natural approach used in the seat on the wooden partition.
↓绚丽色彩的灯光营造出西方浪漫的夜色风情 / Gorgeous lighting create a romantic night out of the West style.

A DESIGN CASE FOR RESTAURANT

高品煲汤肥牛餐厅

15

【坐落地点】长春市宽平大路1411号铭仁大厦；【面积】2 300 m^2

【设计】李文

【主要材料】地砖、铁板、铁管、石膏板、艺术涂料、阻燃塑料管、绳

【摄影】李晓军

餐厅的各区域形式各异而风格统一，浪漫主义层出不穷。客人们来到这里，可以暂时忘却那些扰人的琐事，伴着蒸腾的气泡、靓汤，举起手中的酒杯，大快朵颐，不醉不归。

给人留下深刻印象的是如森林般神奇的就餐大厅。顶棚及过道中间用绿色的绳子随意织成不规则的网，构成视觉的中心。白色的人造大理石桌、斑马纹的布艺椅子，深色的地砖，一明一暗。设计师在这里设计了一个夹层，松木的桌椅与空间隔断以一个整体的形式出现，起伏而流畅，个性强烈的同时又不失整体感。

与“森林就餐大厅”相接的是半开放的包厢区，也是极好的洽谈区域。鲜红色的金属杆像红裙摆一样围绕着上升，在天花上归于一面镜子。这是几个用曲线相互连接形成的好似笼子一般的私密地点。曲线巧妙地演化成隔断，绵延至深处。

与海洋有关的神话故事被融进设计中。在另外一个就餐区，绿色的曲面形石膏板艺术棚顶是海面，墙壁上红色的规则的玻璃板圆形图案以及棚顶垂下来的透明玻璃珠子，都是气泡的衍生体。

Restaurant-style differences in the regional forms of unity, everywhere is a romantic ambience. Past guests have come here, you can temporarily forget about those annoying chores, and accompanied by transpiration of the bubble, soup, and raised the hands of the cup, people very happy.

Dining hall, give left a deep impression. The middle of a green roof and the aisle rope woven into a random irregular networks, create visual center. Artificial white marble tables, zebra-patterned fabric chairs, dark floor tiles, a dark one out. Here designers designed a sandwich, pine tables and chairs and room dividers to form a whole, undulating and smooth, while a strong personality without losing the overall sense.

And the "Forest Dining Hall" connecting the semi-open box area, are also excellent discussion area. Bright red metal bar like a red skirt around the same rise in the ceiling attributed to a mirror. This is the curve of several interconnected with the formation of the general's private place is like a cage. Curve cleverly evolved into partition, stretching from the depths.

Ocean-related myths have been woven into the design. In another dining area, a green art ceiling gypsum board is a sea wall on the rules of the red glass circle designs, as well as the past Ceiling hanging transparent glass beads, are derived from the bubble body.

↑整洁与繁杂的对比中形成用餐区 / Neat contrast with the complex to form dining area.

↑加层空间 / Add layer of space.

↓用绳和网以及金属杆件编织出不同的功能区域 / Rope and nets, and pieces of woven metal rod different functional areas.

↑林中木屋 / Forest huts.
↓平面图 / Plan

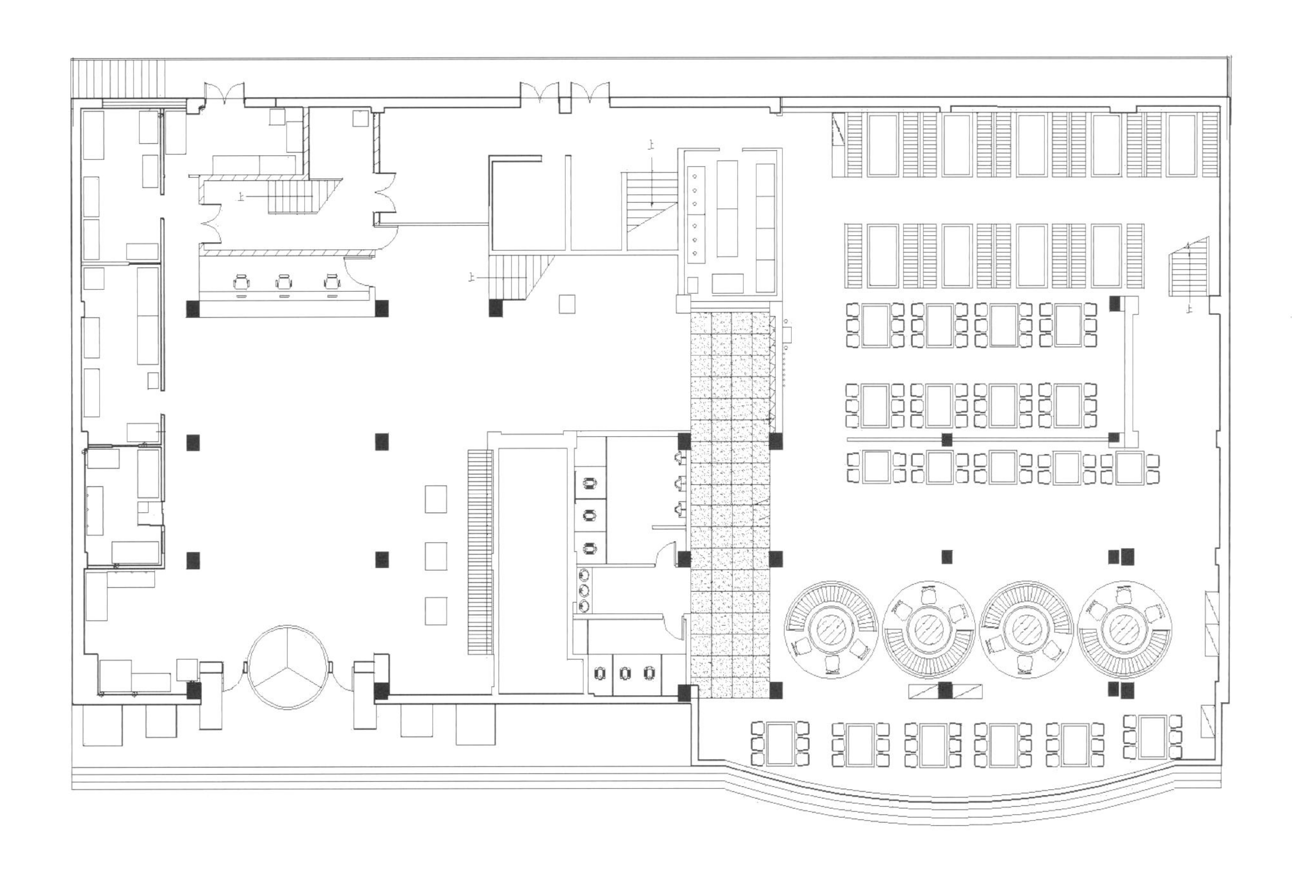

↑气泡 / Bubbles.
←半开放包厢 / Semi-open box.
↓绚丽色彩的灯光 / Brilliant colors of light.；↓金属编织网细部 / Metal braid detail

WES
STYLERES

西式餐厅
TERN
STAURANT

97 RESTAURANT IN SHANGHAI

上海97餐厅

16

【坐落地点】上海复兴公园内；【面积】480 m^2
【设计】Lance Smith
【设计单位】麟研部建筑设计咨询（上海）有限公司
【摄影】申强

整个空间分为两个区域，主用餐区和一侧的贵宾休闲区。两个功能区风格迥然不同，在丰富空间形式的同时，又互有补益。空间的设计起点为餐区中央椭圆形吧台，这个大型的白色吧台一下就将空间氛围调动起来，然后将别致的律动衍射到用餐区。在这个功能区里，材质与氛围的搭配，也是设计师关注的重点，他们选择了米黄色和木色来让环境更加轻松闲适。

紧邻主用餐区的白色休闲区，采用了完全不同的设计主题。与用餐区使用的略显沉静私密的暗色不同，休闲区用色明快，可以看成是用餐区的延伸和补充。在以几何图案为设计要点的休闲区里，白色格纹围挡与沙发相呼应，以花为主题的深色壁纸又与地面的黑白几何纹对应，用色的关联性使得空间杂而不乱，意趣十足，呈现出很强的折中主义风格。休闲区还设有一个夹层阁楼，增加了整个空间的层次感。

由于设计的初衷是要营造出一个精致的私密的用餐环境，但在非用餐时间，又可以使其变成一个高档的休闲会所。所以在空间灵活性的考量上，设计师让所有的桌子都安装电动马达，这样就可以很容易地调整高度，以适应不同要求。相应的，灯光也能随意调节，实现从充满情调用餐环境到幽暗多变会所氛围的轻松转变。

The whole space is divided into two regions, the main dining area and the side of the VIP lounge. Two very different style of functional areas, in the form of abundant space for the same time, but also each other helpful. The design space for the dining area from the central oval-shaped bar, the large white bar room atmosphere for what will be mobilized, and then the rhythms of diffraction to the unique dining area. In this functional area, the materials and atmosphere matching, but also the focus of designers, they chose the color beige and wood to make the environment easier leisure.

The main dining area adjacent to the white leisure area, using a completely different design theme. Dining area with private use of the dark quiet slightly different colors, bright lounge, dining area can be seen as an extension of and complement. In a geometric pattern for the design features of the leisure district, the white sofa cut off and echoes to the dark-colored flowers as the theme of black and white wallpaper again with the ground corresponds to the geometric patterns, colors, makes the relevance of space and clean, showing a strong eclecticism style. Lounge also has a mezzanine loft, an increase of the level of a sense of the whole space.

The design was originally designed to create an intimate fine dining environment, but in the non-meal time, they can make it into a high-end leisure club. From the space, with flexibility in mind, the designer put all the tables are installed in electric motors, so that you can easily adjust the height to suit different requirements. Accordingly, the light can be freely adjusted from a meal full of flavor changing the environment to the dark atmosphere of a relaxed club changes.

↑入口 / Entrance

↑白色椭圆吧台 / White oval bar.
→通往二楼的楼梯 / Leading to the second floor of the escalator.
↓墙面同样呈现波浪造型 / Show the same shape of a wave wall.

↑通往二楼的扶梯 / Leading to the second floor of the escalator；↑纯白空间 / White space.
←休闲区设有一个夹层阁楼 / Recreation area is equipped with a mezzanine loft.
↓休闲区 / Leisure areas.

BRAND RESTAURANT

Brand牛排馆

17

【坐落地点】Monte Carlo酒店；【面积】约687 m^2
【设计】GRAFT设计公司〔德〕
【主要用材】烟熏橡木地板、LED动画板、牛皮图案
【摄影】Ricardo Ridecos

天花板的装饰顶是本案整个空间的亮点。它由多块装饰板组成，本身呈现出别致的类似牛皮质感的图案。不同深度发亮的装饰板营造了美妙绝伦的壮观场景，使客人感觉舒适、宁静。装饰顶由255块嵌板构成，这些嵌板共有8种规格。嵌板上部的空间有MEP组件。

在休息区，地面与天花板的区分变得明显起来，这使客人可以感受到天花板整体的动态美感。透明的可以滑动的原片玻璃闭合时可以作为第二道屏障，同时又可呈现空间的视觉连续性。酒吧是休息区较远一侧垂直表面的焦点，它为Brand牛排馆增添了一个新的维度。

主餐饮区由三个区组成，在该区域内，地面向餐厅的后方缓缓升起，将客人引至主场。各种各样的餐桌布局丰富了人们就餐的美好体验。宴会坐椅上方的铜镜上蚀刻着熠熠生辉、实体动画的树枝，从主餐饮区的墙上延伸出去，使客人可以仔细地观察后面反射的空间，并同时欣赏这片抽象的丛林。

包厢区完全是一个私人世界。客人在此享受一种完全不同的体验。房间内部的磨砂玻璃移门可以进一步划分空间，使得餐厅更具私密性。云母石墙后方、木制餐具橱上方灯具灯光的微妙变化会引起人们对室外景观的无限遐思。

This case the design space is highlighted by a ceiling. It consists of several pieces, decorative plates, showing their own unique texture similar to leather pattern. Shiny decorative panels at different depths to create the scene of the spectacular to make guests feel comfortable, quiet. Decorative Top 255 panels from the composition of these panels a total of eight kinds of specifications, the upper panel space MEP components.

In the rest area, clear the distinction between ground and ceiling, which makes the guests can feel the dynamic beauty of the ceiling as a whole. The original can be a transparent glass slide can be closed as a second barrier, while the visual continuity of the space can also show. Bar is the focal point resting area, which has Brand Steak house added a new dimension.

The main dining area consists of three zones formed in the region, ground-oriented restaurant in the rear slowly rising, will the guests to the casino cited. A variety of dining table enrich people's needs. Banquet chairs at the top of the copper mirror etched with shiny, solid animation branch, from the main dining area extends out of the wall, so that guests can carefully observe the space behind the reflector, and at the same time to enjoy this piece of abstract jungle.

Box area is a completely private world. Guests at this enjoy a completely different experience. Frosted glass sliding door inside the room can be further divided into space, makes the restaurant more privacy. Mica stone wall behind the wooden cutlery cabinet above the lamp lighting can cause subtle changes in people's unlimited outdoor landscape reveries.

↑独特的选材，赋予空间自然的精华 / A unique selection to give the space the essence of nature.

↑入口处精致的LOGO体现了它的风格 / The LOGO at the entrance reflects its exquisite style.
↓换个角度又是不同的场景 / From another point of view is a different feeling.

↑吧台的台面材质犹如冰制 /Bar of the table light Rubing material；↑天花板迷人的图案装饰 / Charming decorative ceiling patterns.
↓丰富和自由的材料 / A rich and free materials.

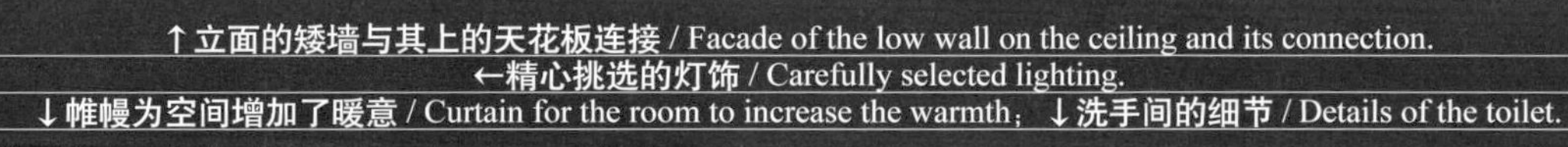
↑立面的矮墙与其上的天花板连接 / Facade of the low wall on the ceiling and its connection.
←精心挑选的灯饰 / Carefully selected lighting.
↓帷幔为空间增加了暖意 / Curtain for the room to increase the warmth；↓洗手间的细节 / Details of the toilet.

BROWN SUGAR
JAZZ CLUB & RESTAURANT
Brown Sugar音乐餐厅

18

【坐落地点】上海市新天地北里广场太仓路；【面积】826 m^2
【设计】甘泰来；【参与设计】高泉瑜、梁世伟；【设计公司】齐物设计
【主要用材】石材、墨镜、银箔、木皮、镀钛、皮革
【摄影】卢震宇

餐厅的入口十分简单和低调，不规则的墙面带出很多不确定感，这些与内里的丰富和华美形成鲜明的对比。设计师利用这种"对比"的策略，塑造一系列多层次的空间感官体验，让空间形式不仅立足于机能(form fallows function)，亦带出新时代中都会餐饮娱乐空间的新观念：形式立足于乐趣(form fallows fun)。

无论是楼梯、舞台的背景墙，还是整个室内空间，不难看出设计师在照明设计上花费了很多精力。简单而言，四个字即可以形容：轻、重、缓、急。轻，是漂浮，如一系列桌上的烛台灯；重，是聚焦，如投射于木皮墙面的聚光灯；缓，是微晕，如阶梯间接照明灯；急，是变幻，如主舞台区的LED舞台灯。这是一种让灯光可以表达情感，亦可和空间形体或物料对话的感性式灯光计划。

空间中除了"新旧融合"剧目的持续上演外，亦透过基地本身具有的挑高属性，加上局部地坪高层变化的安排，让整体空间以"剧场"的概念，塑造出各种不同定位区域(如：主舞台、散桌区、高吧区、沙发区、VIP区等)，且进一步使座位区域间，亦形成"看"与"被看"，互为舞台或观众席的多重关系。设计师在材料的运用上可谓别出心裁，以黑檀木人字形拼花，蔓延于空间中大部分地面以至墙面，塑造温暖复古之调性。

The restaurant's entrance is very simple and low-key, irregular walls with a lot of uncertainty, these with the inside of the rich and gorgeous in stark contrast. Designers can take advantage of this "contrast" strategy, create a series of multi-layered sensory experience of space, so that spatial form is not only form fallows function, also give rise to a new era of dining and entertainment space will be a new concept: form fallows fun.

Whether stairs, stage backdrop, or the entire interior space, can be seen lighting design designers spend a lot of energy. In short, words that can be described as: light, heavy, slow, anxious. Light, is floating, such as a candlestick table lamps; weight, is focused, such as the projection on the veneer wall is in the spotlight; slow, is halo, such as the ladder of indirect lighting; urgent, is changing, such as the main stage area of the LED the stage lights. This is a way for light can express emotion, and space can be physical or emotional material dialogue-style lighting scheme.

Space in addition to "old and new fusion" continuing staged, but also through the base itself has the high-ceilinged attributes, coupled with changes in the arrangement of local strata, so that the overall space to "theater" concept, create all kinds of different location areas, and further inter-regional seats, but also form a "look" and "be seen" each other the relationship between multi-stage or auditorium. Designer on the use of the material can be described as innovative, in the space on the ground most of the design of the word ebony mosaic, to create a warm and classic blend.

↑接待区色调暗沉，不规则的墙面带出很多不确定感 / Color dull reception area, irregular walls with a lot of uncertainty.

↑餐厅入口，建筑的印记与新鲜的时尚元素相结合 / Restaurant entrance, the architectural imprint and fresh combination of fashion elements.

↑阶梯间接照明灯，是空间的一抹亮色 / Step Indirect lighting is bright touch of the space.

↓墙面采用黑檀木的拼花塑造出温暖复古的感觉 / The walls of the parquet with Ebony create a warm retro feeling.

↑主舞台背景墙上的LED灯 / Background on the wall of the main stage LED light.
↓二层平面图 / 2nd Floor Plan.

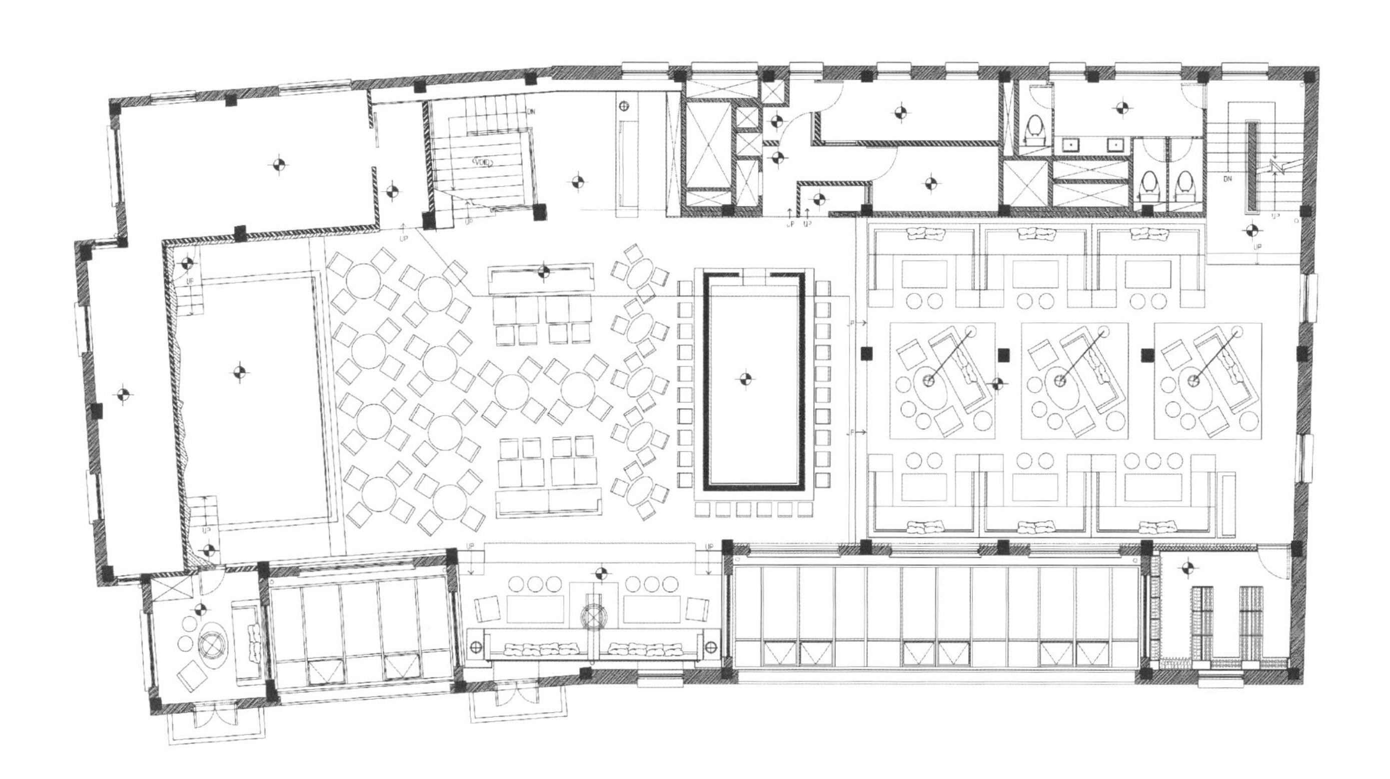

↑空间的过道，与外立面的砖瓦质感一脉相承 / Aisle space and the facade of the brick surface of the same.

↑红色聚光灯投射于木皮的墙面上 / Red spotlight projection on the veneer of the wall.

↓沙发区是极为独特的一隅，它的沉静与舞台上的喧嚣形成对比 / Sofa area is a very unique corner, it's quiet and the stage in contrast to the din.

↑光在吊顶桁架结构中渲染，突显出老建筑的美 / Light in the ceiling truss structure, rendering, highlights the beauty of old buildings.

↓三层平面图 / 3rd floor plan.

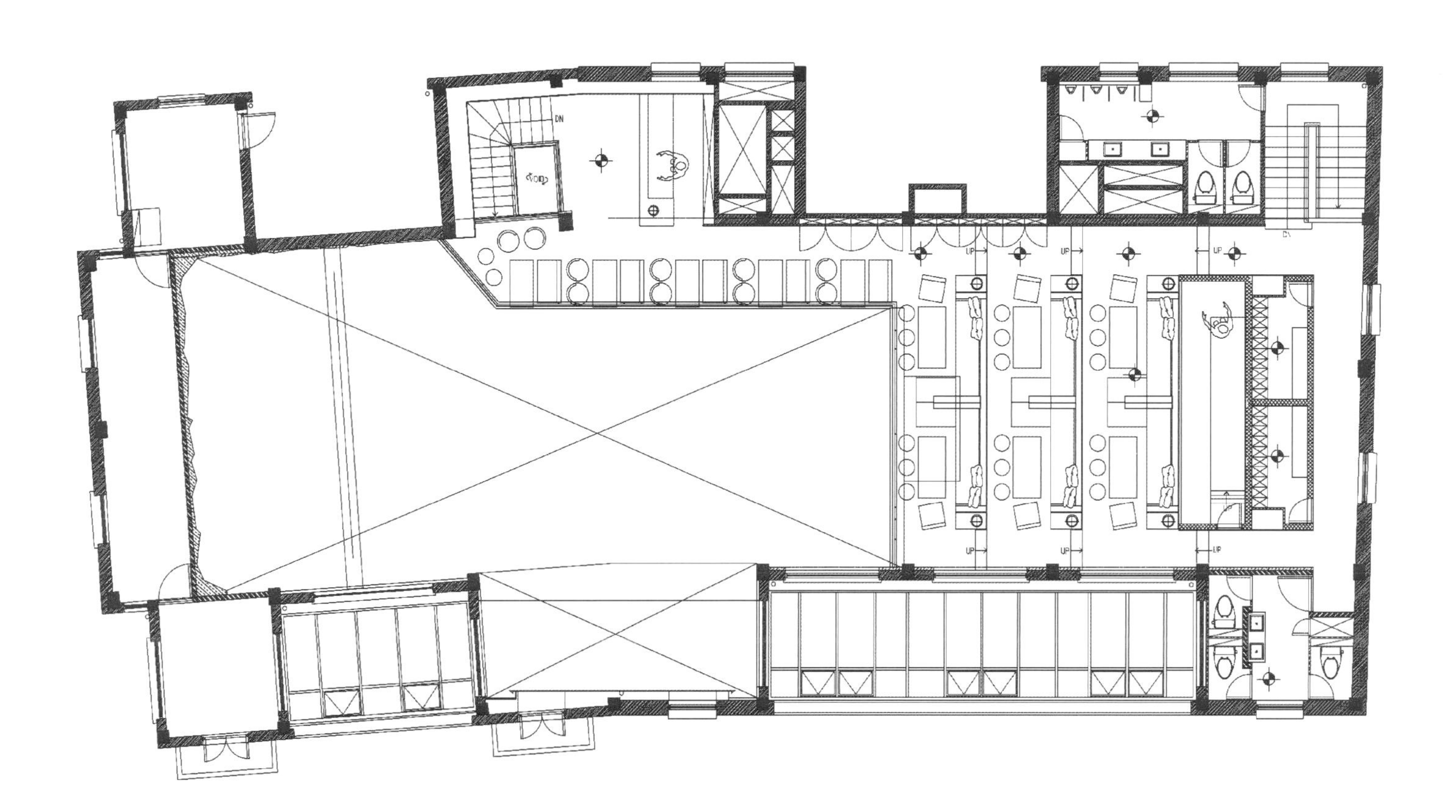

METOO CAFE RESTAURANT

蜜桃西餐厅

19

【坐落地点】杭州金华路丝联166创意园
【面积】约420 m^2
【设计】张健；【设计公司】齐物设计
【主要用材】旧木材、旧砖、旧家具

沿着厚实的、红砖铺设的小路就可以进入"METOO CAFE"室内了。白色的木结构从旧有的墙面延伸出来，与透明的玻璃完美结合，塑造了一个童话般剔透敞亮的阳光房。在这里，每一扇门都是敞开的，犹如一位年轻漂亮的女主人微笑着迎接客人们的到来。

设计师说，他在设计这个地方的时候用了一种很特别的手法——拼凑。就是利用一些旧材料、回收家具与原有厂房遗留的物件进行衔接，使其融合、协调，然后产生情调。厂房的旧砖墙被刷成了白色，与进门处白色的木构架色彩一致，让空间的主调更加典雅恬淡。主就餐区小小的陈设台上摆放着各处搜集而来旧时的物品，那些旧电器、上世纪七八十年代人们熟悉的玩具，呼应着空间的主题。陈设桌上方，横向长条的镜子，在视觉上扩大了空间。

蜜桃的其中一道墙的中部被挖出了一个圆形的孔洞，长长的桌子延伸过去，给客人别样的会餐体验。卫生间的外部看似简陋，可是推门而入也能让人有惊艳的感觉。偌大的落地窗用在了这里，外界青翠欲滴的植物就好像一帧巨幅的风景照片，让人有空间转换的错觉。四个VIP包厢看似相同，却还是在细节上体现了差异，可供客人享受自己的私人空间不被打扰。

Along the thick, red brick paved paths can enter the "METOO CAFE". A white wooden structure extending out from the original walls, with the perfect combination of transparent glass, created a fairy-tale light and spacious sun room. Here, every door is open, like a beautiful young mistress with a smile to greet the arrival of guests.

Designer said he was in the design of this place would use a very special way - a variety of mix. Is the use of some old materials, old furniture and things left over from the original plant converge to integration, coordination, and then create mood. The old brick factory building has been painted white, with the door at the white line, so that the main theme of a more elegant and quiet room. The main dining area furnishings collected from around the desks are old items, the old electrical appliances, the familiar toys of the last century, corresponding to the space design. Display table at the top, stood horizontal mirror, visually expand the space.

METOO CAFE in which the central part of a wall has been dug a circular hole, table extended in the past, giving guests a special dining experience. Bathroom outside looks primitive, but the push to gain access can also surprise people are feeling. Used here, the huge windows, the outside world like a green plant a huge landscape photographs, giving people the illusion of space conversion. 4 VIP box looks the same, but still reflects the differences in detail are available for guests to enjoy their own private space is not disturbed.

↑接待区色调暗沉，不规则的墙面带出很多不确定感 / Circular holes, an extension of the long table in the past, different kind of dinner guests to experience.

↑沿着厚实的、红砖铺设的小路走进室内 / Along the thick, red brick paved path into the interior.
↓VIP包厢 / VIP box.

↑主就餐区的陈设台上摆放着各处搜集而来旧时的物品 / The main dining area furnishings desks are collected from old items.

↓平面图 / Plan

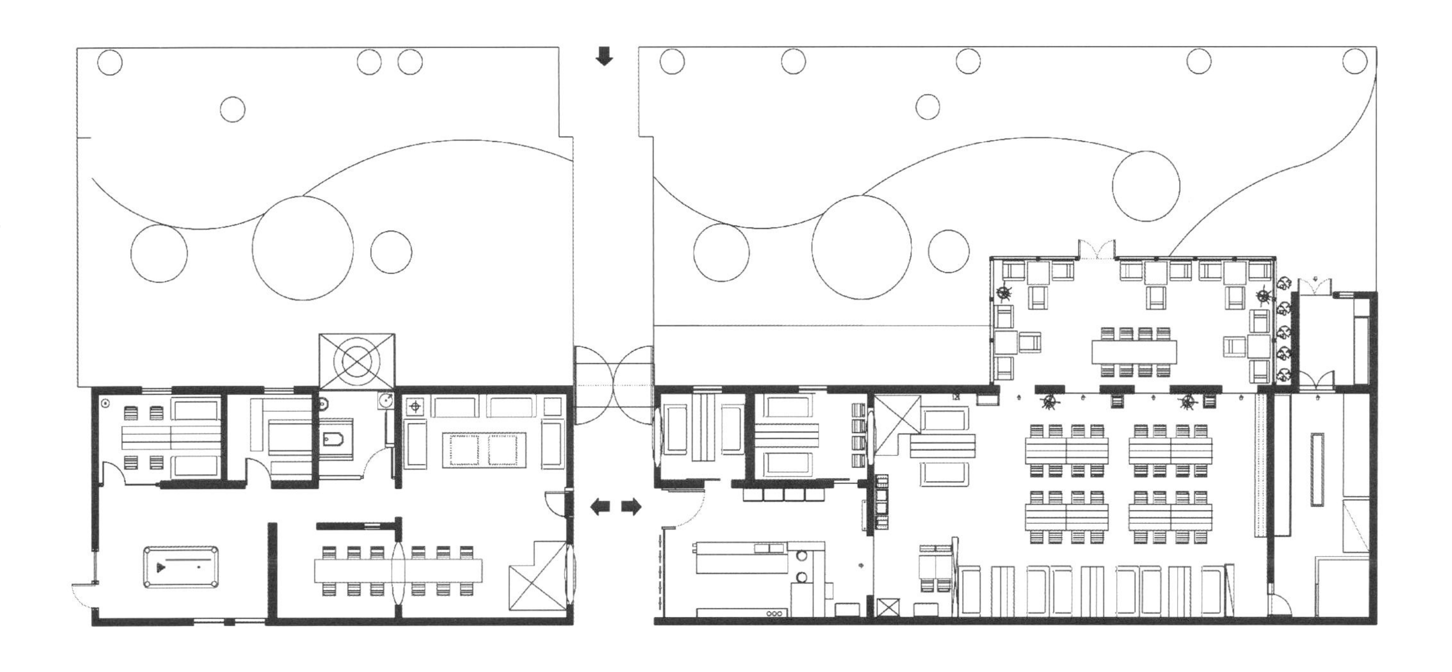

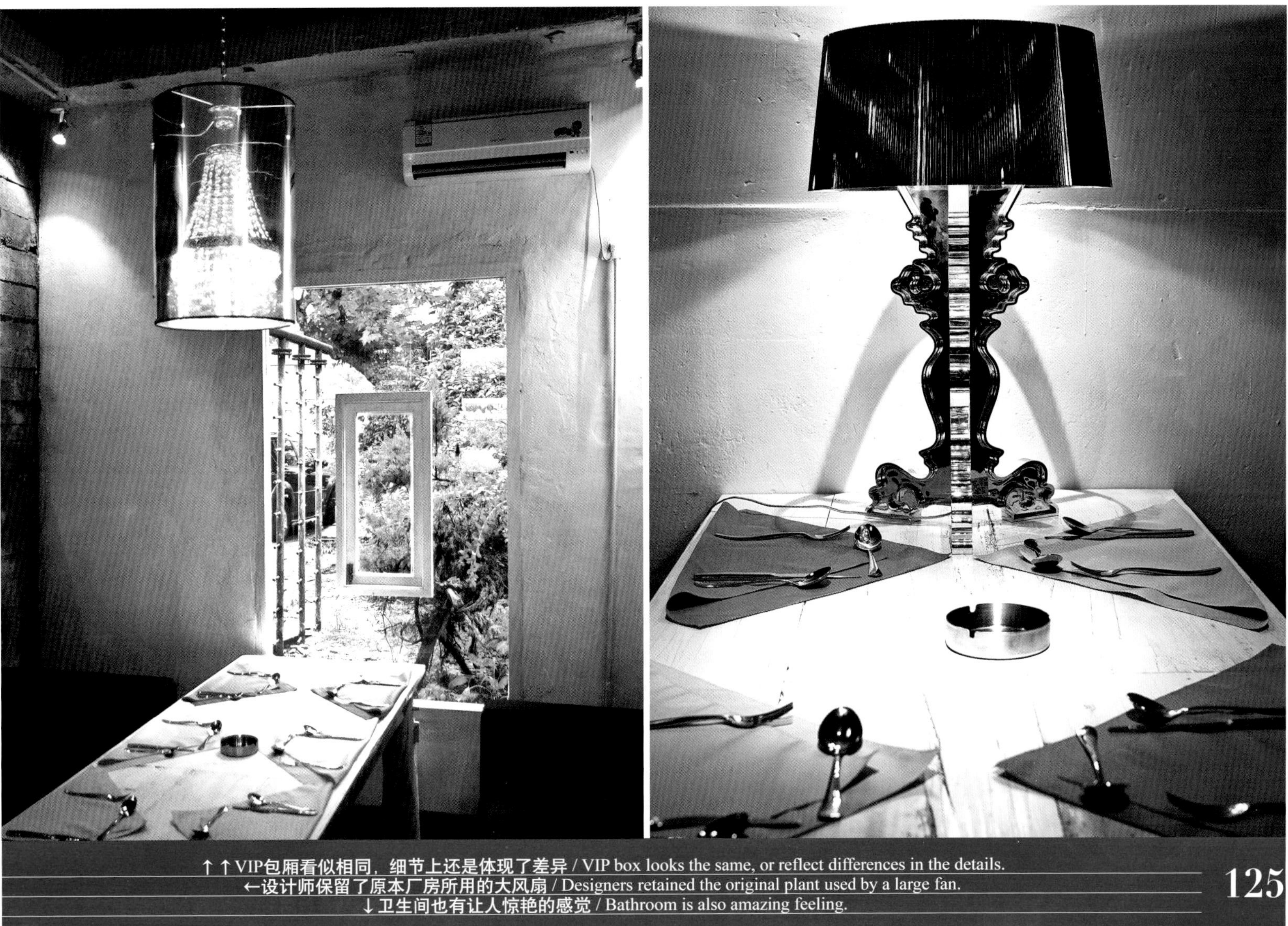

↑↑VIP包厢看似相同，细节上还是体现了差异 / VIP box looks the same, or reflect differences in the details.
←设计师保留了原本厂房所用的大风扇 / Designers retained the original plant used by a large fan.
↓卫生间也有让人惊艳的感觉 / Bathroom is also amazing feeling.

ECHO ITALIAN RESTAURANT

艾可意大利餐厅

20

【坐落地点】北京海淀区远大路25号

【面积】284 m²

【设计】香港KWSG

【摄影】贾方

在本案餐厅里，设计师将几何元素流畅地运用，橙红色的坐椅慵懒地躺在餐厅的大厅里，顾客一进去就被"火热"的意大利氛围包裹住；造型简洁有如未来太空舱的过道设计，时刻充满着时尚的Lounge Music，更不禁令人有时空地域错置的惊愕；而与大厅融为一体的隔断设计，又为整个餐厅带来一份异国的安静。餐厅空间的每一处设计都是别具心思的，透着淡淡木纹的木质天花，低调华丽的银色马赛克墙壁，红、绿、黄相间的布艺沙发、桌椅，打造出意大利热情洋溢的波西米亚风情。墙上金属边框里的壁画，灯下的水晶挂帘，搁板上随意的红酒瓶，又勾勒出了意大利式的闲适恬美。

后现代主义是餐厅造型的本源。地面图案带有如拂晓之时岁月留在苍岩上的形状，同时又是参观者的无形指引；泛着幽蓝银色光彩的马赛克墙面，轻盈和粗壮的结合；镜面式的空间隔断，提供线性形状与雕塑体块动和静、轻和重的对比；充满后现代感的体块组合体现现代建筑的某种灵动精神却又不失传统民族风格。一切是这样和谐，热烈与安逸、美食与音乐……

一切从空间出发，一切从体验出发。空间即美，空间即精神，空间即企业形象。意大利餐厅不仅仅只是美食消闲的地方，它更是一个放松心身的场所，你会就像迷恋上Tiramisu一样，迷恋上这里。

In this case restaurants, designers will be fluent in geometric elements, orange and red seats lying on the lobby restaurant, the customer was inside a "hot" atmosphere of parcels to live in Italy; modeling simple passages, always full of fashion The Lounge Music, more people can not help but have time territorial disorder surprise; integrated with the hall of the partition design, but also for the entire restaurant to bring in a foreign country quiet. Restaurant space, creative designs are everywhere, and reveals a pale wooden ceilings, ornate silver mosaic walls, red, green, yellow and white fabric sofa, tables, chairs, Bohemian style has been built out. Metal frame inside the wall murals, crystal lamp curtain, free of red wine bottles on the shelf, but also outlines the Italian-style leisure and comfort.

Postmodernism is a restaurant shaped origin. Ground pattern with the shape of the rock, while the invisible guides visitors; suffused with dark blue silver glorious mosaic walls, a combination of light and thick; mirror-style room dividers, to provide a linear shape and sculpture, body mass action and quiet, light and heavy contrast; filled with a combination of post-modernist reflect a certain spirit of modern architecture yet lose national tradition. Everything is so harmonious, warm and comfortable, food and music ... ...

Everything starting from space and experience.Space is beautiful, rich in the spirit of space, space is a kind of corporate image. Gourmet Italian restaurant is not just a place for leisure, it is a psychosomatic place to relax, you will like the obsession on the Tiramisu, like the obsession here.

↑热烈风情的意大利 / A warm Italian style.

CARRUADES
DE LAFITE
2000
CARRUADES
DE LAFITE
2002
SAINT
CHEVALIER

←↑金属壁画、水晶挂帘，勾勒出意大利式的舒适 / Metal murals, crystal Gualian, depicts Italian-style comfort.
↑银色马赛克墙壁让过路空间变得丰富起来 / Silver mosaic wall for crossing the room to become enriched.
↓平面图 / Plan

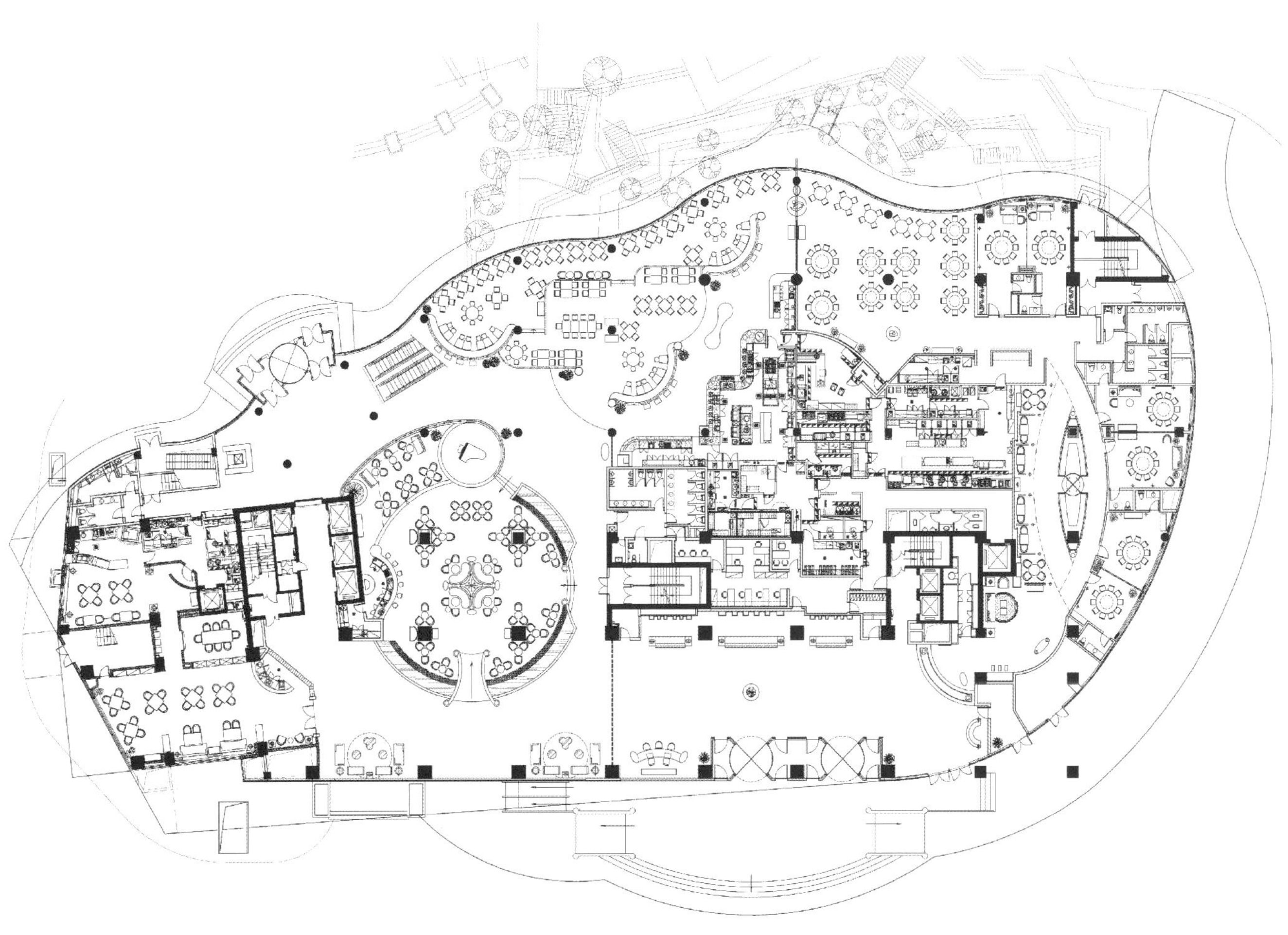

MR PIZZA'S RESTAURANT

米斯特比萨天津店

21

【坐落地点】天津南门外大街天地烩金茂广场；【面积】一层 100 m^2，二层 480 m^2

【设计】Yeonhee Keum〔韩〕、宛艳玲

【主要材料】定做雕刻地砖、红色玻璃、白色乳胶漆

【摄影】贾方

作为一个富有活力的西餐品牌，Mr Pizza天津南市店的设计主题也围绕着童心、想像、活力、现代而展开。餐厅整体设计基调为红色和白色，希望突出活力、热情。

从店外透过玻璃就可以窥见一楼入口处孩子们的游玩空间。这样可以隐约窥见里面的结构，比起整面的玻璃墙，显得更有情趣。

二楼"红色蛋形座位"的右边是往洗手间的通道。一般来说，顾客不喜欢这样的位置。设计师用隔断尽量挡住顾客的视线，做出与等候区域的墙结构相连接的墙体，墙壁中间还特意追加了一个壁灯使气氛更加温馨。这样设计使得这个位置比其他位置更具吸引力、更受欢迎。楼梯旁的位置也总是不太受欢迎，于是设计师利用曲线型的装饰隔断挡住视线，避免顾客产生不安感，并且特意为座位安排了Saarinen椅子，突出其特殊性。

设计师在餐厅的天花板上用了红白两色和椭圆形的概念做装饰。二楼餐厅中除了楼梯旁边的Saarinen椅与众不同之外，还有一部分座位使用了与餐厅不一样的红色panton椅，成为视觉亮点，并用彩色玻璃来完成装饰隔断。

As a dynamic Western brands, Mr Pizza designed around the innocence, imagination, vitality, modern initiated. Restaurant overall design for the red and white, want to highlight the vitality and passion.

From the outside the shop can get a glimpse through the glass floor at the entrance to the children's play space. This vaguely visible inside the structure, compared to the entire surface of the glass walls, seems more interesting. Arc wall itself, using special-shaped, the wall thickness of each location is different, even in the vertical direction as well, with the curve to make a curved look.

On the second floor of "red egg-shaped seat" is the right of access to the toilet. Generally speaking, customers do not like this place. Designers try to block the customer with the cut off line of sight, making the walls of the structure and the waiting area connected to the wall, the wall also adds an intermediate wall to make atmosphere more warm. This design makes this position more attractive than other locations and more popular. The location near the stairs is not always popular, so designers to use curved decorative partition block line of sight, to avoid customer created a sense of insecurity, and specially for the seating arrangement in the Saarinen chair, a prominent position.

Designers used the restaurant's ceiling color and oval-shaped red and white two concepts as decoration. In addition to the stairs on the second floor restaurant next to the Saarinen chairs unique, there are some seats with the restaurant not the same as using the red panton chairs, a vision center and decorated with stained glass to complete the cut off.

↑热烈风情的意大利 / A warm Italian style.

↑走进入口门，会看到孩子们的游玩空间 / Into the entrance gate, you will see the children's play space.
↓红色墙壁和Saarinen 椅 / The red walls and Saarinen chairs.

↑顶部的天花 / At the top of the ceiling；↑弧形墙的不同角度 / Curved walls of the different angles.
↓一层平面图 / Floor plan.

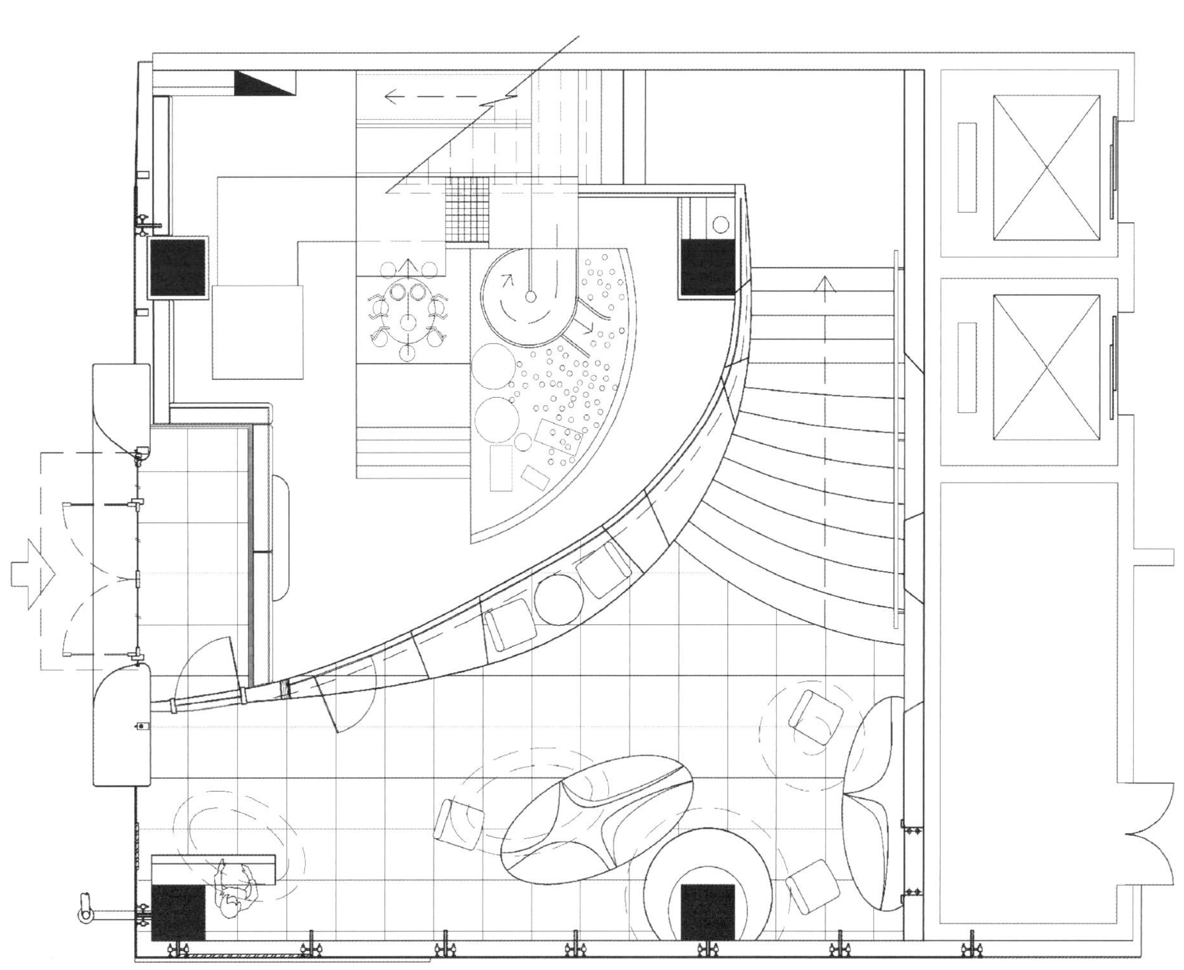

↑装饰隔断部分使用人造石打磨成蛋形 / Decorative cut off the use of artificial stone grinding into the egg-shaped.

↓弧形天花勾勒出的空间 / Curved ceiling outlines space.

↑用彩色玻璃装饰的隔断 / Decorated with stained glass partition；↑奢华气氛的卫生间 / Luxury bathroom atmosphere.
↓二层平面图 / 2nd floor plan.

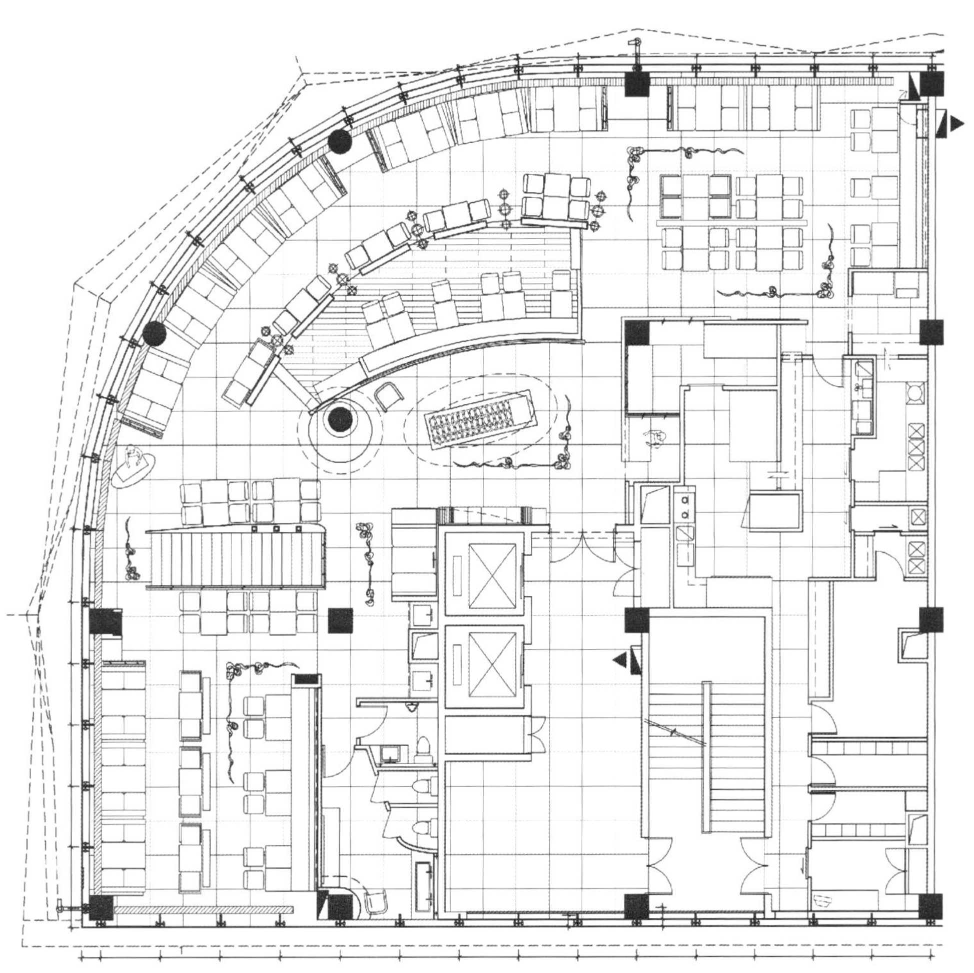

BOULUD PALACE FRENCH RESTAURANT

布鲁宫法国餐厅

22

【坐落地点】北京前门23号项目内
【面积】约760m²
【设计】Gilles & Boissier
【设计单位】Gilles & Boissier设计公司

餐厅内部设计的灵感源自20世纪经典的"大沙龙"（Grand Salons）时期的室内装饰风格，同时也从万花筒对色彩的折射作用中得到启发，通过特殊材质的应用来展现餐厅异彩纷呈的精致和优雅。为尊崇建筑本身的历史意义，Gilles & Boissier在其内部的设计上也尽量保留了建筑原有的庄重的风格。

餐厅的色调搭配巧妙，与一些独特的装饰完美结合，既传达出对历史和建筑的尊重，又散发着一种现代的优雅气息。进入建筑内部，上等的胡桃木材质的、装饰房间墙壁精致的细木护壁板随处可见。

豪华的餐厅是洽谈和举办重要的餐会的地方。沿着墙面，白色的细木护壁板用松木制成，黑色木质结构在墙面上铺展开来展示着现代的荣光，由17世纪法国画家Poussin创作的巨幅画制，与精致的镜子和法国古典灯具交相辉映，别样精彩。混合着现代风格和英式风格的家具随处可见，非常的舒适。家具上覆盖的是带有浓浓法兰西风情的特色织物。

来到粉红餐室，客人的感受又一次转换。精心布置的陈设、用做旧的粉红织物和粉红皮革制成的家具，唯美而经典。这不可思议的比例和奢华的材质让这个房间的气质更加的个性化以及女性化。休闲室和酒吧比邻而居。休闲室可以用做气氛轻松的午餐或其他聚会，而酒吧的空间延续超现实主义的风格。VIP房间更是风格迥异，令人赞叹。

Restaurant interior design inspiration from the 20th century, the classic "Grand Salons" period style interior decoration, but also from a kaleidoscope of color inspired by the refraction of the active, through the application of special materials to showcase the restaurant's sophisticated and elegant. To showcase the historical significance of building itself, designers in the design of its internal work to preserve the original architectural style of solemn.

The restaurant's color with cleverly decorated with some of the unique combination. Not only convey the respect of history and architecture, but also exudes a modern atmosphere. Get inside the buildings, high-quality walnut and exquisite plates everywhere.

Luxury restaurant is important to discuss and organize dinner place. Along the walls, the fine white wood baseboard is made with pine, black wooden structures in the wall capped started to show the glory of modern, from the 17th century French painter Poussin paintings created by a huge system, with the exquisite mirror and French classic lamps match up extremely well. There are many restaurant mixed with a modern style and British style furniture. Furniture is covered with a thick French-style characteristics of the fabric.

Into a special pink dining room, guests feel another conversion. Well-arranged furnishings, used the old pink fabric and pink leather furniture, the classical aesthetic. This is incredible proportions and luxurious materials make this room a more personalized temperament and feminine. Lounge and bar next door. Lounge can be used as a relaxed lunch, or other gatherings, while the continuation of the space bar of realism style. VIP room is also different styles, is impressive.

↑白色的帷幔让空间更显浪漫 / A white curtain so that space is even more romantic.

↑镜子的使用增大了空间的尺度 / The use of the mirror increases the scale of space.
↓入口立面图 / Entrance elevation.

↑细节中体现餐厅的尊贵奢华 / Reflected in the details of the restaurant's exclusive luxury.
↓一层平面图 / Floor plan.

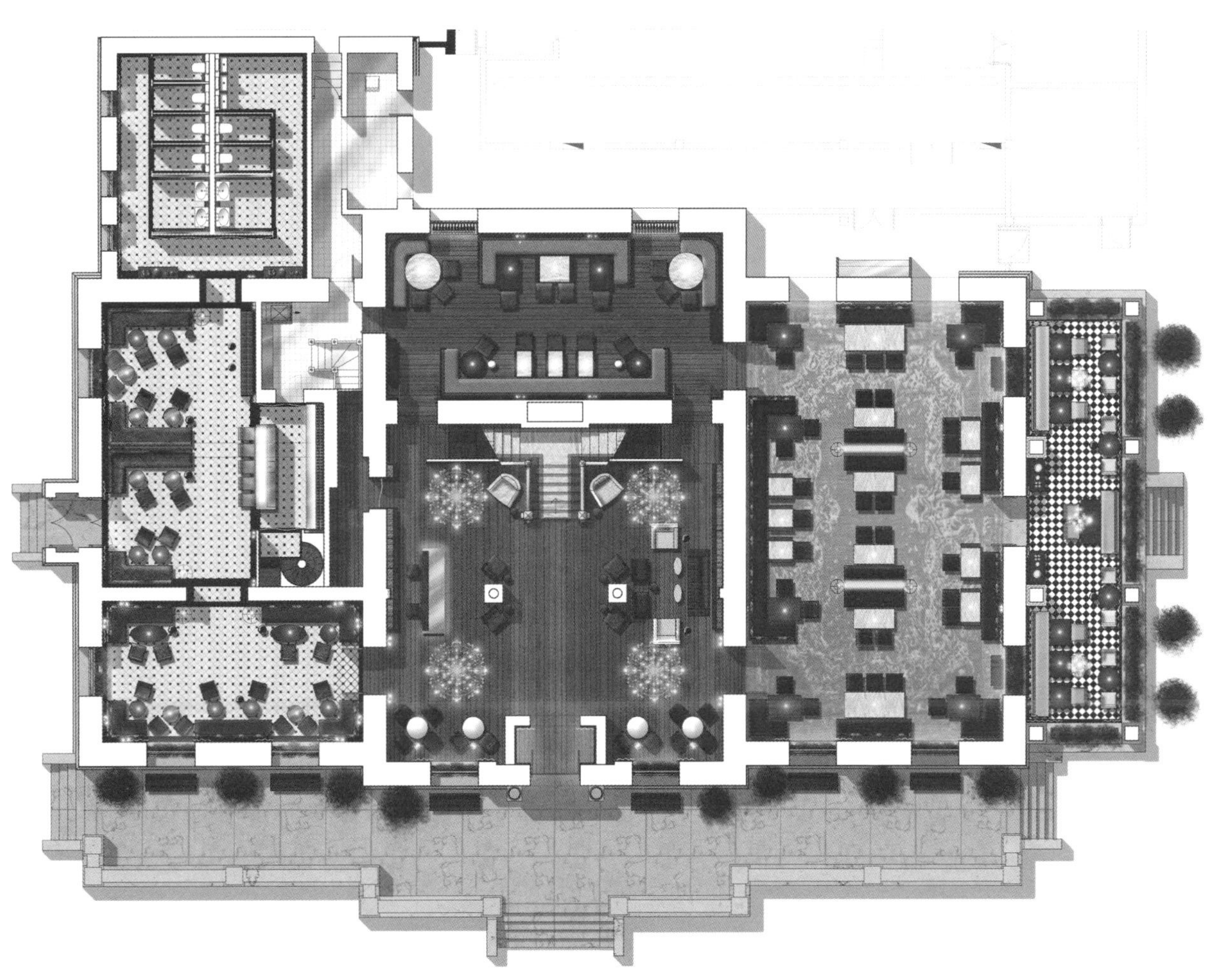

↑法国画家Poussin创作的巨幅画制 / French painter Poussin painting system created huge.
↓大厅立面图 / Hall elevation.

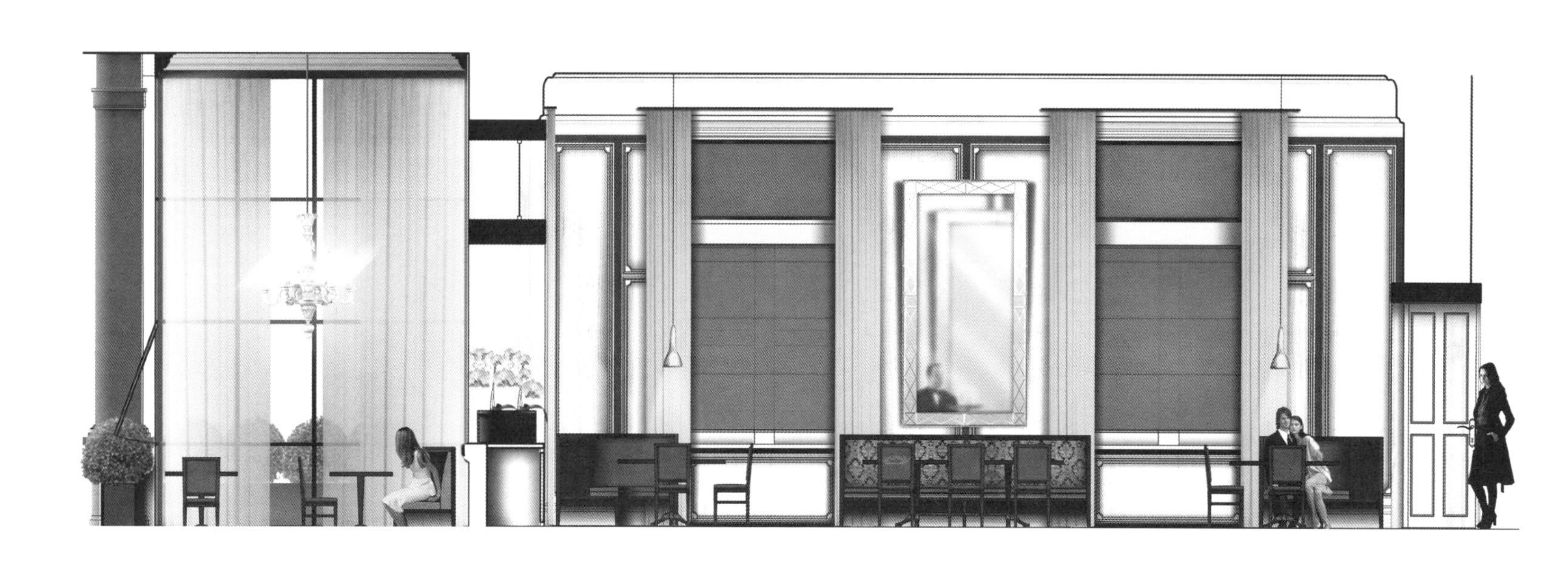

↑↑白色的帷幔让空间更显浪漫 / A white curtain so that space is even more romantic.
↓粉红餐室焕发出的与众不同的戏剧舞台般的效果 / Dining out distinctive pink glow as the effect of the stage drama.

↑楼梯处，原建筑的柱子和楼梯被保留 / Stairs at the original construction of the columns and staircases have been retained.
↓楼梯处立面图 / Stairs at elevation.

↑ VIP房风格迥异 / VIP rooms in different styles and external.
↓二层平面图 / 2nd floor plan.

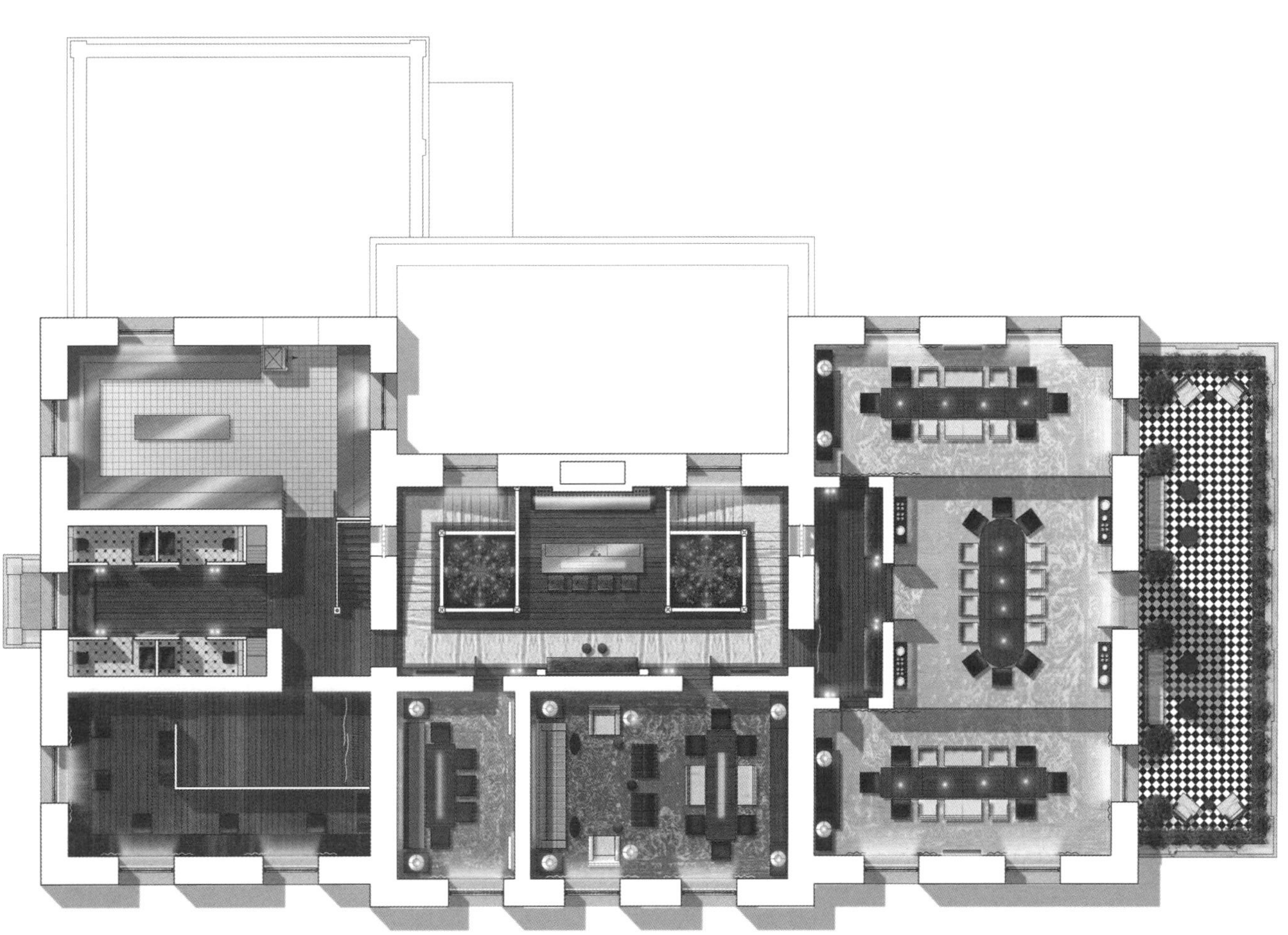

NOBU IN DUBAI

Nobu餐厅

23

【坐落地点】迪拜棕榈岛亚特兰蒂斯酒店
【面积】11 500平方英尺
【设计】Rockwell Group〔美〕
【摄影】Eric Laignel

餐厅入口处的酒吧有一条60英尺长的环形石墙，墙上用激光雕刻了许多盛开的樱桃花形象。酒吧墙面用的是3英寸厚的木板，地面则是用的玛瑙石做的装饰。大厅的设计延续了酒吧的风格，装饰了许多半透明的樱桃花图案，放大的图案呈一个椭圆形排列在私密区的上方。

整个就餐区被蕉麻叶子编织起来的三维空间环绕着，给客人一种置身海浪的感觉。颜色各异的蕉麻叶子像瀑布一样从天而降，包裹着这个富有特色的空间。

寿司吧和这个具有强烈视觉冲击感的就餐大厅并列。空间的地面采用了传统的亚洲风格，黑色的水磨石地面镶嵌的是一个一个大小不一的竹子的横截面。而墙面则用了白色作为主背景色，同样镶嵌了竹材的圆形横截面做装饰。白色背景的一边用蓝色的磨砂镜面过渡到服务台，而另一边则用装饰有金属花朵图案的弯曲墙面连接包间。

木质的拱门在包间门口迎接着各位来客。包间的墙壁上，一面还是用传统的花朵图案做装饰，另一面则用整齐排列的各种酒瓶来装点。而头顶下垂的灯箱上，镂空的花朵图案正在空间上方偷偷盛开。

Restaurant bar at the entrance there is a 60-foot-long circular stone wall, the wall with a laser engraving a lot of the image of cherry blossoms in full bloom. Bar wall using a 3-inch thick wood, the ground is done using agate stone decoration. Hall's design follows the style of the bar, decorated with a number of translucent cherry flower patterns, enlarge the pattern was arranged in an oval above the private area.

Throughout the dining areas are woven abaca leaves are three-dimensional space around them, giving the guests a feeling of exposure to the waves. Abaca leaves of varying colors, like the same waterfall from heaven, wrapped in this distinctive space.

Sushi Bar and this has a strong sense of visual impact dining hall side by side. The space on the ground using traditional Asian style, a black terrazzo floor is a mosaic of a cross-section of bamboo of different sizes. White walls are used as the main background color, the same mosaic of circular cross-section of bamboo for decoration. One side of a white background with blue frosted mirror the transition to a desk, while the other side with decorative flower pattern curved metal wall connecting rooms

Wooden arch at the entrance to greet the rooms our guests. Packets between the walls, one side is the traditional flower design as decoration on one side and neatly arranged with a variety of bottles to decorate. The top of the head drooping of the light box, the hollow space at the top of the flower pattern are secretly in full bloom.

↑餐厅的墙壁和天花都用巨大的焦麻叶子装饰起来 / The restaurant's walls and ceilings are a huge coke with hemp leaves decorated.

↑传统的寿司吧 / A traditional sushi bar.

↓酒吧的吊灯是传统的中东风格 / The chandelier is a traditional pub in the style of the Middle East.

↑ 中东风格的吊灯 / Middle East-style chandelier.

↑ 地面和墙面都镶嵌了竹子做的装饰 / Mosaic floors and walls are made of bamboo decoration.

↓拱形的木块和弯曲的蕉麻叶子，增加了空间的运动感 / Arched wood and curved abaca leaves, increased the sense of spatial movement.

↑→木板墙面增加了寿司吧的原汁原味 / Wooden wall to increase the true face of the sushi bar.
↓包厢里整齐的酒柜 / The wine cooler box was neat.

ISSIMO RESTAURANT

Issimo餐厅

24

【坐落地点】上海南京西路930号
【面积】1 200 m²
【设计】Darryl Goveas
【摄影】吴永长

整个餐厅由三个主要部分组成：入口处的酒吧区、主用膳区和两个大小不一的私人用膳包房。设计师把不同的文化背景融到室内环境中，使得设计不仅充满当代艺术的气息，还造就着浓厚的乡情，散发着独特的魅力。

主餐厅宽敞而丰富，首先映入眼帘的是一条长约12m的橄榄木吧台。这个吧台不规则的外轮廓、纯粹的木质感，带来一种自然的感觉，另外稍显亮丽的颜色也使得这个正餐厅变得活泼很多。吧台一侧是正餐区，另一侧则是开放式厨房。

主餐厅的设计以简约为主，但为了配合历史建筑独特的魅力，建筑内采用了一些与之相呼应的元素：家具基本都选择色彩感较为沉重的颜色，如设计精致的木地板、棕色的桌椅、操作台等；木地板的铺装方式也与历史建筑复杂的线脚取得某种方面的相似感。

餐厅后方设有两个单独的包房餐室，为晚宴或酒会提供灵活的座位，并能调整空间大小形成3个区域。餐厅更设置了可从天花拉下设计独特的布帘，以划分个别空间。

Restaurant consists of three main components: the entrance of the bar area, main dining area and two private dining private rooms. Designers of different cultural backgrounds melt into the indoor environment, making space for contemporary art is not only full of flavor, there is a strong nostalgia, exudes a unique charm.

Main Restaurant spacious and rich, the first thing that catches your eyes is a 12m long bar of olive wood. This counter irregular outer contours, pure wood texture to bring a natural feel, while the bright color makes this is a lot of restaurants have become active. Bar at the side of a meals area, on the other side is the open kitchen.

The main restaurant's design simplicity mainly, in order to cope with the characteristics of historic buildings, buildings used some complementary elements: the basic furniture, choose colors more heavy sense of color, such as the design exquisite wood floors, brown tables and chairs, the operation Taiwan, etc.; wooden floor of the pavement patterns and the complex of historic buildings to achieve a certain line of legs similar to a sense of respect. In addition, the walls black and white theme with trees and paintings and are from New York, the famous artist and photographer Michael Weber's hand, which is even more for the dining area adds a unique ambience.

Restaurant There are two separate private dining room behind the dining room for dinner or a cocktail reception to provide a flexible seating, and can adjust the size of the formation of three regions of space. Restaurant is designed to pull down the curtains from the ceiling to divide the individual space.

↑镶满钻石切割式镜子 / Diamond-studded mirror.

↑楼梯踢面使用红棕色条纹状的木地板 / Use the stairs kicking surface reddish-brown striped wood flooring.
↓楼梯的扶手展示出古朴与现代融合 / Staircase handrails demonstrate ancient and modern fusion.

↑↓ 设计师将酒吧区域与走道设计分开，使酒吧成为一个独立的空间 /
Designers will be designed to separate the bar area with the walkway, so that the bar became a separate space.

↑ 长约12 m的橄榄木吧台的不规则的外轮廓 / About 12 m of the olive wood bar outside of the irregular contours.
← 私人用膳区 / A private dining area.
↓ 主要餐厅区 / The main restaurant area.

BUDDAKAN RESTAURANT

Buddakan餐厅

25

【坐落地点】Meat Packing District 75 9th AVE，纽约
【设计】Gilles & Boissier；【设计主持】Philippe Starck,Dorothée Boissier
【设计单位】Gilles et Boissier工作室
【摄影】Eric Laigne

餐厅设计的灵感源自对东方庙宇特色的现代诠释，通过不同材质及空间布局的应用，来展现餐厅异彩纷呈的精致和优雅。设计中的色调搭配巧妙，与一些极具代表的东方文化特征的元素完美结合，既传达出对东方文化的尊重，又散发出一种时尚现代的优雅气息。

进入接待区，一场视觉的盛宴由此展开。接待区与街道位于同一标高，通过几级踏步，四扇高大的橡木门，提示空间的转换，划分出休息室和接待区。

休息室的装饰风格继续沿用了跨文化的组合。由Liaigre设计的方形的橡木餐桌和条状的长凳简洁大方，不失东方特色，但椅面上包裹着的却是西方的天鹅绒织锦。整体布局上借鉴了典型的中国徽州民居厅堂的布局方式。

沿休息室一侧的木楼梯拾级而下，来到主宴会厅，空间随高度转化，从东方的宁静含蓄步入西方的热情奔放。巴洛克风格的主宴会厅金碧辉煌，传达出热烈、明快的气氛，将整个餐厅的空间节奏烘托到了高潮。

地下室的餐室运用了佛龛的概念，并被布置得舒适异常。与之毗邻区域则以"图书馆"为主题，用高大的书架围合出一个相对私密的就餐区域，周围环绕的书架上陈列着仿制的经书，封面全部为金色。

Restaurant inspired by the design characteristics of the modern interpretation of an oriental temple, through the different materials and space layout, to show the restaurant's sophisticated and elegant. Design colors with clever, very representative of the East with some elements of the perfect combination of cultural identity, both to convey respect for Eastern culture, but also exudes a stylish atmosphere of modern elegance.

Into the reception area, a visual feast hence commenced. The reception area and the street at the same elevation, through the steps, appeared four tall oak doors - prompted the conversion of space, and carved out lounge and reception area.

Lounge decoration continue to be used in cross-cultural style. Designed by Liaigre square strips of oak tables and benches simple and generous, without losing the Oriental features, but the chairs are wrapped surface of the West, velvet brocade.

Lounge along the side of the wooden stairs down, came to the main banquet hall, room with a high degree of conversion, implicit from the east into the West, the quiet passionate. Baroque-style gilded the main banquet hall to convey a warm, lively atmosphere, the whole dining room for the rhythm contrast to a climax.

The use of the basement dining room of the shrines of the concept, and was furnished comfortable exception. And areas are next to the "Library" is the theme here, with tall bookshelves enclosure out of a relatively intimate dining area, surrounded by bookshelves display the imitation of the classics, covers all of the gold.

↑宴会厅中央的主餐桌由橡木制成，有30英尺长 / The main banquet hall of the Central dining table made from oak, 30 feet long.

↑黑檀色的橡木门上，用时尚、现代手法演绎着中国古典元素 / In the black oak door, with a modern interpretation of a classical Chinese elements techniques.

↓穿过橡木大门，看见的是一幅17世纪的巴洛克风格的油画 / Through the oak door to see is a 17th century Baroque style of painting.

↓休息室一角 / Lounge corner.

↑休息室的布局借鉴了徽州民居厅堂的布局方式 / Lounge layout draws the layout of Huizhou houses hall way.
↓ 透过色彩鲜艳的花格窗能看到位于下层的主餐室 / Through the colorful flower in the lower lattice windows to see the main dining room.

↑ 西方的油画和东方的经文之间的碰撞 / Western oil painting and the contrast between the Eastern scriptures.

← 豪华的宴会厅 / Luxurious banquet hall.

↓ 四周墙面上木制的书架围合出以图书馆为主题的就餐区域 / Wooden shelves on the wall around the enclosure out of the library-themed dining area.

PARKSIDE RESTAURANT

Parkside餐厅

26

【坐落地点】上海市卢湾区

【面积】150 m²

【设计】赵牧桓；【建筑设计】牧桓建筑+灯光设计顾问

【摄影】牧桓建筑+灯光设计顾问 周宇贤

在本案中，设计师决定在项目中融入上海的历史文化背景。餐厅坐落于一个有着70年历史的老式公寓一层，与大多数老上海的洋房一样，这里面或多或少都存留着历史的痕迹。设计师刻意保留了部分原始建筑形态。如黄铜管被暴露在外，以非对称的方式存在，形成了3D的效果，成为一种有趣的天花装饰物。墙上的老砖被刷上一层白漆，仿佛一伸手就能触到一幅幅古老的画面。地面的图案与天花有异曲同工之妙，在浓郁的老上海气息中又蕴涵着现代几何学的韵律感。

在预算不高的前提之下，我们采用了温和的现代主义手法，试图诠释的是巴洛克风格奢华的精神内核。作为21世纪的东方设计师，我们将不断探索的是一种与当今时代相辅相成的设计方式。

In this case, the designers decided to integrate into the project, Shanghai's historical and cultural background. Restaurant is located in an old apartment with a layer of the history of the seventies, with most of the old house in Shanghai, like the inside is more or less retained the traces of history. Designers deliberately retained some of the original architectural form. Such as the brass is exposed to the existence of non-symmetrical manner, forming a 3D effect, become an interesting ceiling decoration. The old brick wall was a layer of white paint on the brush, as if one could reach out touch screen depicting the ancient. The ground pattern is similar with the ceiling and wonderful, and in the rich flavor of old Shanghai, he also implies a sense of rhythm of modern geometry.

In the budget, under the premise is not high, designers used a moderate modernist tactics in an effort to explain the spirit of the Baroque-style luxury connotations. As the 21st century, the East designers, we will continue to explore with the contemporary era is a complementary design approach.

↑餐厅位于老式公寓的一层 / Restaurant is located in a layer of old-fashioned apartment.

↑墙上的装饰均以阵列的方式排开 / Wall decorations are arranged at an array of ways.
↑天花上的黄铜管是对原始的保留 / Brass on the ceiling is the original reservation.
↓餐厅空间的纵深很狭长 / Restaurant space is very long and narrow depth.

↑墙面上的装饰 / Decoration on the wall；↑ 中心过道区 / Center aisle area.
↓ 平面图 / Plan

HAKKASAN ISTANBUL

Hakkasan餐厅

27

【坐落地点】伊斯坦布尔，土耳其
【面积】1 800 m²
【设计】Gilles & Boissier〔法〕；【灯光设计】Arnold Chan，from Isometrix
【摄影】Birgitta Wolfgang Drejer, Sister Agency

本案坐落于在富有传奇色彩的土耳其首都伊斯坦布尔。Hakkasan餐厅所有的室内设计包括陈设都是Gilles & Boissier一手包办的，设计师在入口处设计了木制的架子，上面摆放着气质高贵的蝴蝶兰，白色的花朵安静的绽放。架子的下面还放置了漂亮的香薰用的器皿，全方位的彰显餐厅的设计理念。

半开敞性质的厨房就设在那些蓝色的玻璃之后，透过玻璃可以隐约看到里面厨师们忙碌的身影，想像里面热火朝天的景象。服务台在细节上透露出高级餐厅的精致，木条凸凹有致的排列，并且精心勾勒出花纹。在长廊的木架结构上，设计师陈列了一些白色的小灯，渲染这些安静的氛围。中式的窗花和门框在这里随处可见，这也是Gilles & Boissier最爱的设计元素之一。长廊的尽头，突出的吧台设计引人注目，花瓶造型的装饰灯洒下旖旎的光线，渲染了动人的情调；抽象的石质背景墙前各色名酒荟萃。在大堂的周围还设有一些较为私密的空间，围合的部分运用了金色的装饰墙，墙上的云纹精致华丽，让这里的氛围更加浓厚纯粹，它被亲切的称作"玲玲厅"。卫生间的设计较其他空间则简洁明快了很多，白色石材搭配具有古典色彩的镜子干净整洁，别具一格。

餐厅的室内设计沿袭了Gilles & Boissier的一贯风格，他们将现代设计理念提炼出来，将其与异域风情的各色木雕装饰及漂亮的壁画完美结合，使得餐厅呈现了别样的神秘优雅的东方情调。

The case is located in the legendary capital of Istanbul, Turkey. Hakkasan restaurant's interior design and furnishings are all personally designed by Gilles & Boissier, designers at the entrance designed wooden shelves, the top decorated with noble qualities of butterfly orchids, white flowers blooming quiet. Placed below the shelf is also a beautiful fragrance used in containers, fully demonstrate the restaurant's design philosophy.

After the blue glass is half open kitchen, you can vaguely see through the glass inside the chefs were busy, imagine that inside the bustling scene. Desk in the details reveals the exquisite restaurants, and orderly arrangement of wood, and carefully outlines the pattern. In the gallery's wooden structure, the designers display a number of small white lights, rendering a quiet atmosphere. Chinese window grilles and door frames to be seen everywhere here, this is Gilles & Boissier, one of favorite design elements.Bar design in the gallery's immediate, vase-shaped decorative lamp reveals a light, rendering a moving mood; stone before the backdrop displays colored wine. Around in the lobby also has a number of private space, enclosed part of the application of the golden decoration of walls, the walls are exquisite ornate patterns, these walls, was the kind known as "Ling-Ling Hall." Bathroom design than the other space is a simple and neat, white stone with a mirror with a classic color clean and unique.

The restaurant's interior design follows the Gilles & Boissier's consistent style, they derived the modern design concepts will be with the exotic combination of wood decoration and murals, making restaurant presents a different kind of mysterious and elegant Oriental flavor.

↑走廊两旁别致的小陈设品使得空间生动起来 / Corridor space furnishings made to life.

↑吧台的设计引人注目，花瓶造型的装饰灯渲染出动人的情调 / Bar design eye-catching, vase-shaped lights rendered pleasing ambience.
←中式风格的窗格花纹增添了东方气质 / Chinese-style pane pattern adds Oriental temperament.
↓ 平面图 / Plan

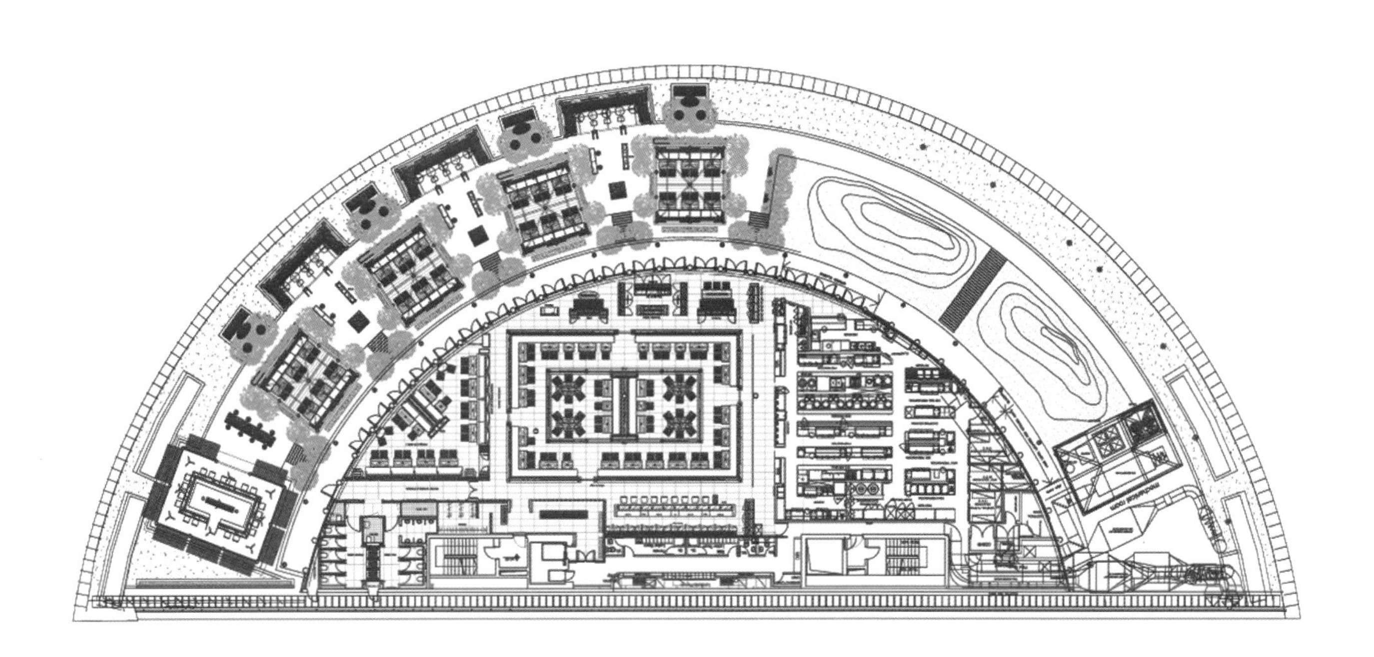

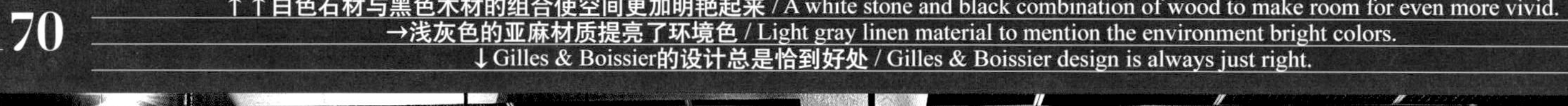

↑↑白色石材与黑色木材的组合使空间更加明艳起来 / A white stone and black combination of wood to make room for even more vivid.
→浅灰色的亚麻材质提亮了环境色 / Light gray linen material to mention the environment bright colors.
↓Gilles & Boissier的设计总是恰到好处 / Gilles & Boissier design is always just right.

↑↓整体的设计不会给人多余的感觉 / The overall design does not give extra feeling.
←泛着黄金色泽的壁画让空间更显华丽 / Reddish gold color of the murals so that space for even more luxury.
↓卫生间的设计相对简洁 / Bathroom design is relatively simple.

PISSARRO RESTAURANT

Pissarro餐厅

28

【坐落地点】香港中环威灵顿街LOOP 11楼
【面积】1 600 平方英尺（149 m^2）
【设计】Michael Young 〔英〕
【摄影】Courtesy of Harlim Djauhar Winata

香港的中环是个不夜天，而Michael就在这里设计了这座法国餐厅，为这个喧闹的市区提供一处宁静优雅的地方用餐。Pissarro Dining 总面积1600 平方英尺，可以同时容纳50人。现居于香港的Michael，近年来跟大陆多间不同的厂商交涉，因此为他带来更多物质材料应用上的灵感，令这间充满现代感的法国餐厅更具特色。

餐厅的大门是Michael和台湾设计工程师 Ken O´ Rouke合作的制成品。他以CNC技术把花纹刻到铝板上，再电镀成设计师心目中的颜色，效果和质量绝对比得上航空标准。墙和天花本身比较薄，于是Michael采用了3M的Di-Noc物料来制造深度和立体感。由于业主要求餐厅以画天空著称的法国画家Pissarro的名字命名，于是设计师也以Pissarro的画作蓝本，再用数码技术把它转化成独一无二的折纸艺术品。为了平衡这些折纸的硬朗感，Michael特地去了富士山一趟，亲手制造出一些玻璃灯罩。餐厅当中的家具则由印度尼西亚的制造商Accupunto提供，当中的Coen椅子（获英国Wallpaper*杂志提名 Best Dining Chair 奖项）更是用柚木和皮革制成的特别版。

这是设计师精心设计的室内空间，是巧妙地把传统手工艺和先进科技都展现出来的佼佼者。

Michael designed this French restaurant in Hong Kong, Central, restaurant is a bustling urban area is located in a quiet and elegant dining areas. Pissarro Dining covers 1,600 square feet, can accommodate 50 people. Currently living in Hong Kong Michael, in recent years, the mainland to negotiate a number of different vendors, so bring more substance and material for his application of the inspiration to make this contemporary French restaurant filled with more features.

The restaurant's door is Michael and Taiwan Design Engineer Ken O'Rouke collaboration products. He took advantage of CNC technology to pattern engraved into the aluminum plate on, then electroplating into the eyes of designer colors, effects and quality comparable to aviation standards. Walls and ceilings in itself is thin, so he used 3M's Di-Noc materials to create depth and dimension. Because restaurant owners are known for painting the sky, named after the French painter Pissarro, so designers also Pissarro's paintings the blueprint for re-use of digital technology to transform it into a unique work of art. To balance the hardness of these origami, Michael specifically went to the Mount Fuji with his own hands to create some of the glass shade. Among restaurant furniture manufacturers from Indonesia Accupunto provided Coen among the chairs is made of teak and leather.

This is a well-designed interior space designer, is artfully traditional handicraft and modern technology were revealed by the leader.

↑酒柜是照明设计是重中之重 / Wine cooler is the lighting design is the key.

↑门板的雏形，已经电镀上了颜色 / Door prototype has been plated on the color.
↓餐厅的Coen椅子 / Coen restaurant chairs.

↑餐厅墙面的装饰亮点 / Restaurant wall decoration highlights.

↓ 平面图 / Plan

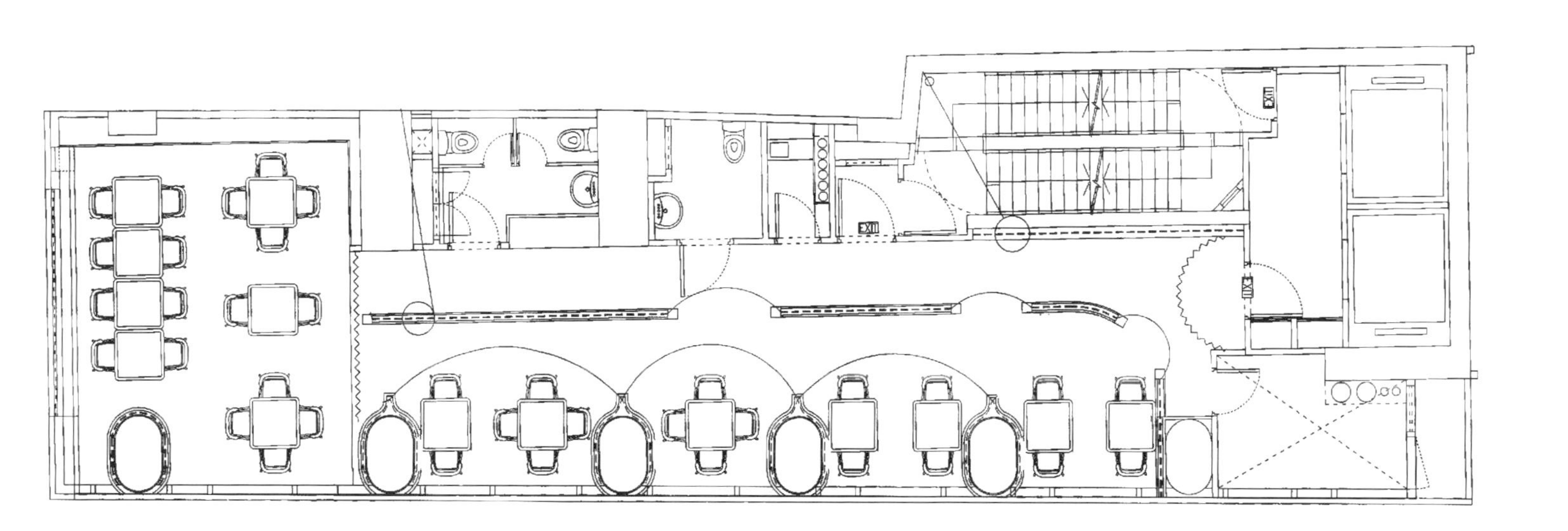

UMINOSACHI LAN SHENG STORE

海之幸兰生店

29

【坐落地点】上海市淮海中路2~8号兰生大厦401室；【面积】560 m²
【设计】小川训央〔日〕、沟胁毅〔日〕
【设计单位】上海英菲柯斯设计咨询有限公司
【摄影】贾方

设计者用最原始自然的原木家具和木结构装饰作为整个设计理念的承载，在“木头”的表现手法上，可以说是点、线、面的集合体。这里没有日式餐厅中司空见惯了的厚重感，设计者运用几何学形态要素以及单纯的点线面结合的手法，尽量避免了因同种材料被反复利用而带来的枯燥感。在不同位置启用了不同的造型手法，意在排除过于繁复的装饰痕迹，从而制造多变的视觉效果，使空间具有简洁明朗的轻快感。从入口的栅格，中庭卡座的隔断屏风到酒水吧台上的吊顶，“木头”被切割成各种形状、被施于各式工艺处理，它时而温婉、时而带有凛凛的质感，自然美丽的木纹清晰可见。

设计者还专门利用激光雕刻的技术，在食肆的墙面、立柱上刻下了无数樱花图案，那犹如雪花般飞舞的花瓣图案、那充满了色彩变化的喷涂墙壁，让原本呆板单调的墙体多了几分柔美与平和。

明明已是初秋，在暮春傍晚的清香里体味着美食，鲜活的食材如同一个水一般温柔的女子，让每个食者在秀色可餐与可餐秀色之间，萌生出朦胧却爽朗的感触，点点滴滴在舌尖温婉地绽放，门外偶尔飘进的樱花雨洒落在你的肩头，邀你与春天共舞。

Designers with the most primitive natural wood furniture and wooden structures decorated to carry the entire design concept of the "wood" and it means, it can be said to be point, line and surface collection. There are no Japanese-style restaurant in the thick sense of the designer to use elements of geometric patterns, as well as a simple points and lines to cover the whole way to avoid the result of repeated use of the same kinds of materials were brought a sense of boring. In different locations using different modeling approach is intended to exclude traces of the decoration is too complex to create varied visual effects, so that space a brighter light with a simple pleasure. From the entrance of the grid, to the entire interior space, "wood" is cut into various shapes, are used everywhere, it is sometimes gentle, sometimes with the stern of the texture, natural beauty of wood grain clearly visible.

Designer also specifies the use of laser engraving technology, in the walls and pillars carved numerous cherry pattern on it like snowflakes fluttering petal pattern, which is full of color change, spraying the walls, so that the original dull monotony of the wall on the increase beautiful and peaceful.

Obviously it is early autumn, in the late spring evening fragrance inside Taste food, fresh ingredients as a water generally gentle woman, so that each food and cuisine in the beautiful scenery between the initiation of the feeling of a hazy but hearty, in the tongue temperature wan to blossom, cherry blossoms floated outside the occasional rain floating down on your shoulders, invite you to the spring dance.

↑入口位置格栅和铁板烧吧台 / Entry location bar and teppanyaki grill.

↑酒水吧台 / Wine bar.

↓入口空间中的格栅 / Entrance space in the grid.

↑铁板烧吧台 / Teppanyaki bar.
↓平面图 / Plan

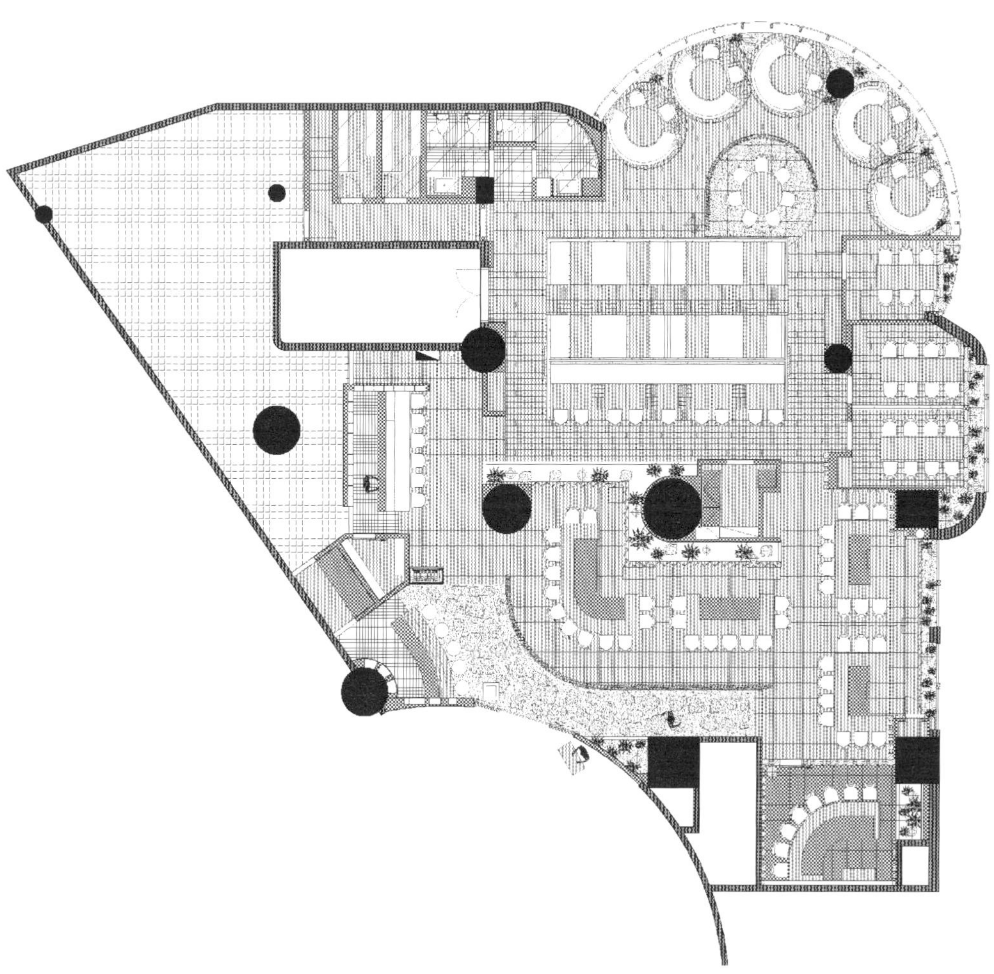

↑中庭空间中的卡座 / Atrium space in the seat.
↓座位上方的木窗 / Seat at the top of the wood.

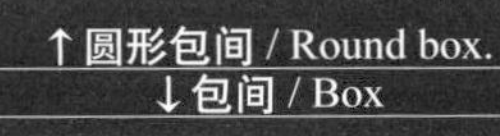

↑圆形包间 / Round box.
↓包间 / Box

A RESTAURANT IN TAIPEI

“翼”餐厅

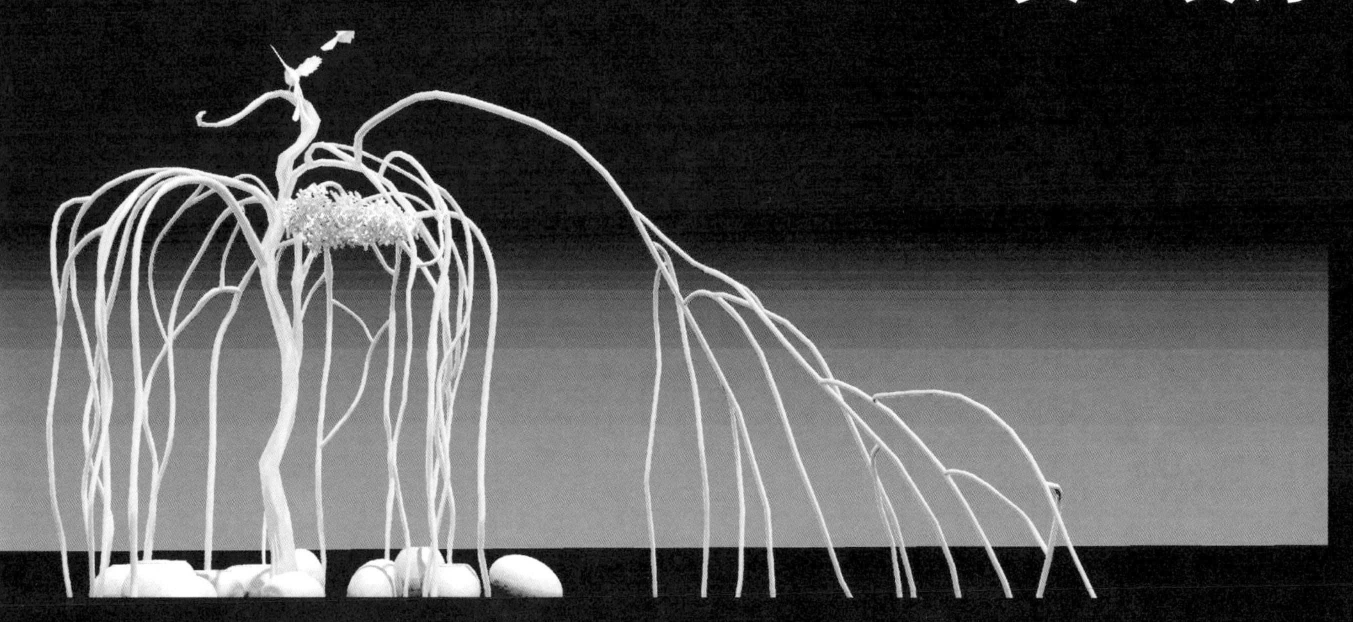

30

【坐落地点】中国台湾台北桃园；【面积】一楼382.8 m²，二楼389.4 m²
【设计】谭精忠；【参与设计】许思宇、何芸妮；【设计公司】动象国际室内装修有限公司
【主要材料】石材有印度黑、火山绿、米黄洞石、印度黑金、黑纹石等
【摄影】ABS color

本餐厅是一间日式餐厅，在这个空间中，我们以一株由大树演化、蔓延盘生出的鸟笼为起点，为人们讲述了一个“飞鸟与树”的故事。艺术赋予空间新的意念，空间与艺术在这里上演了一场精彩的结合。

一层用餐区空间挑高4.8 m的，整体空间均用低彩度沉稳色系营造，为食客提供一进门即感受一份宁静与舒适的用餐氛围。设计上共布局为包厢区、板前料理区和散桌区，以满足不同人数与类型的食客的用餐需求。其中包厢区的设计上被层叠窄板分割的石材大墙包覆着，各包厢呈现连续性独立的空间表现；板前料理区，日式料理的菜肴重点表现在食材的新鲜与精美，考虑到为业主提供一处可灵活调度运用的场所，整面墙就是一幅山形意象画作，呈现出空间的气度。

一楼主入口进门的右侧为连接至二层用餐区的行进动线，整组楼梯墙面装点着以铁板层叠层次错落呈现之艺术创作。二层均为包厢区，设计上与一楼空间的区别采用暖色系。二层前区空间作为进入包厢前的转折区，也可利用做不干扰用餐进行的谈话空间。包厢廊道以细长条型分割风化木作表现处理，期许以细腻的设计手法，营造空间的静腻与长条廊道连续性的层次感。

The restaurant is a Japanese-style restaurant in this space, we evolved a cage by a tree as a starting point for people to talk about a "bird and the tree" story. Art gives room for new ideas, space and art in here, staged a wonderful combination.

Floor space for high-ceilinged dining area 4.8 m, the whole color space used to create a flat calm for the diners to create a quiet and comfortable dining atmosphere. The design of the layout for the box area, board table before the cooking area and scattered areas, in order to meet the different numbers and types of diners dining needs. Box area in which the design has been stacked plates separated narrow stone walls covered with, the boxes show the continuity of an independent space for performance; plate before the cooking area, Japanese-style dishes highlighted by the fresh and fine ingredients, taking into account the owners can be flexibility in the use of a place, the entire wall is a painting Yamagata images, showing a tolerance space.

A layer of the main entrance door to the right of the road line, the whole group in order to iron staircase, the walls decorated with cascading levels of scattered show of artistic creation.The second floor balcony area are designed with the difference between the first floor space with warm system. Anterior two-story space as a turning point into the box before the area can also be used as a conversation space. Slender bar box corridor to the performance of segmentation weathered wooden handle, expectation, by the delicate design techniques to create a space for static greasy and long corridors of the level of a sense of continuity.

↑一株由大树演化、蔓延盘生出的鸟笼 / An evolution from the trees, the spread of the bird cage birth plate.

↑ 一楼散座区 / On the first floor lobby area.
←料理台将新鲜的食材展现出来 / Cooking sets out to show the fresh ingredients.
↓一层平面图 / 1st floor plan.

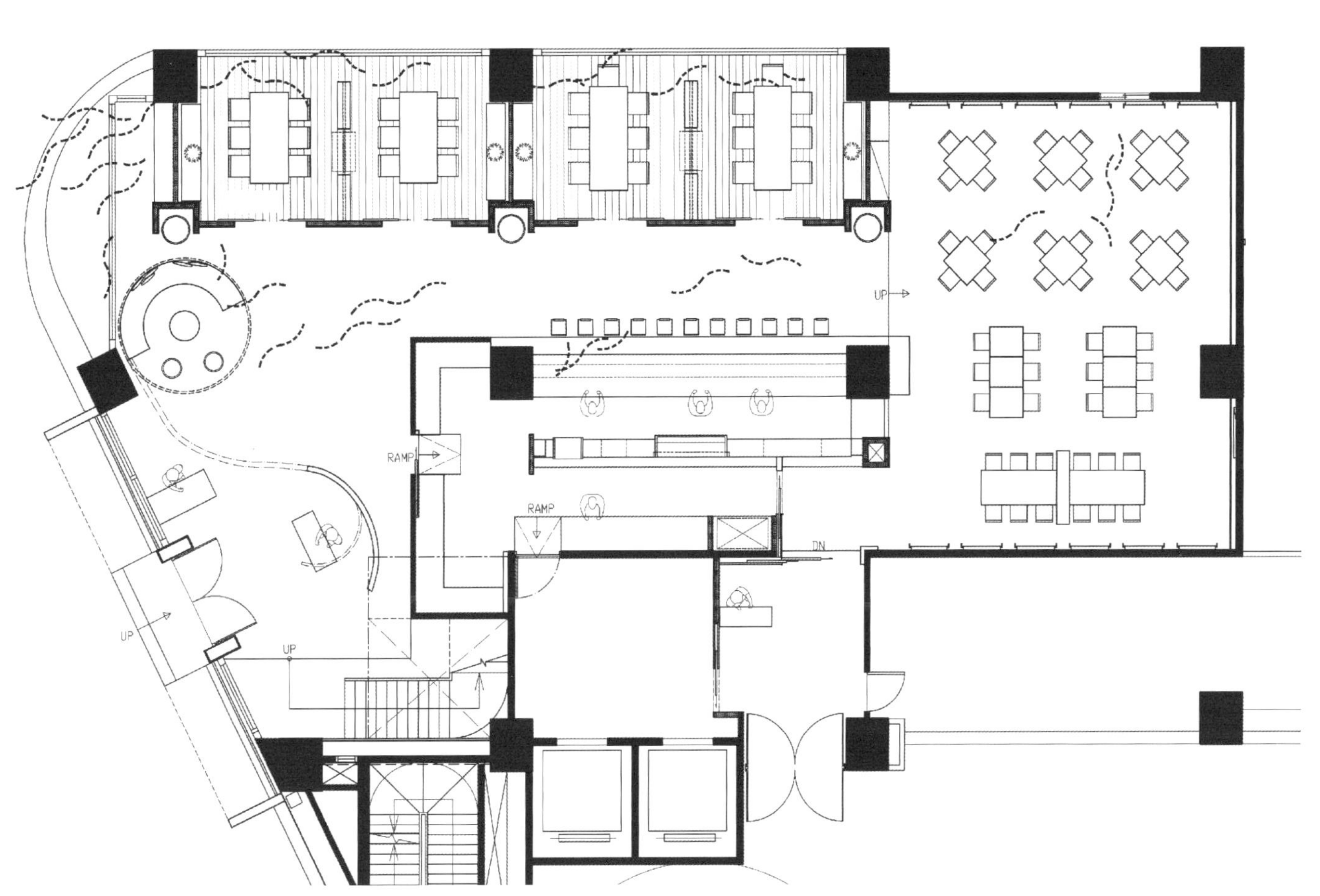

↑二层的楼梯 / The second floor staircase.
↓艺术小品 / Works of art.

↑二层设计上以温暖木质色系为主 / The second floor is designed with warm wood color-based.
↓二层平面图 / 2nd Floor Plan.

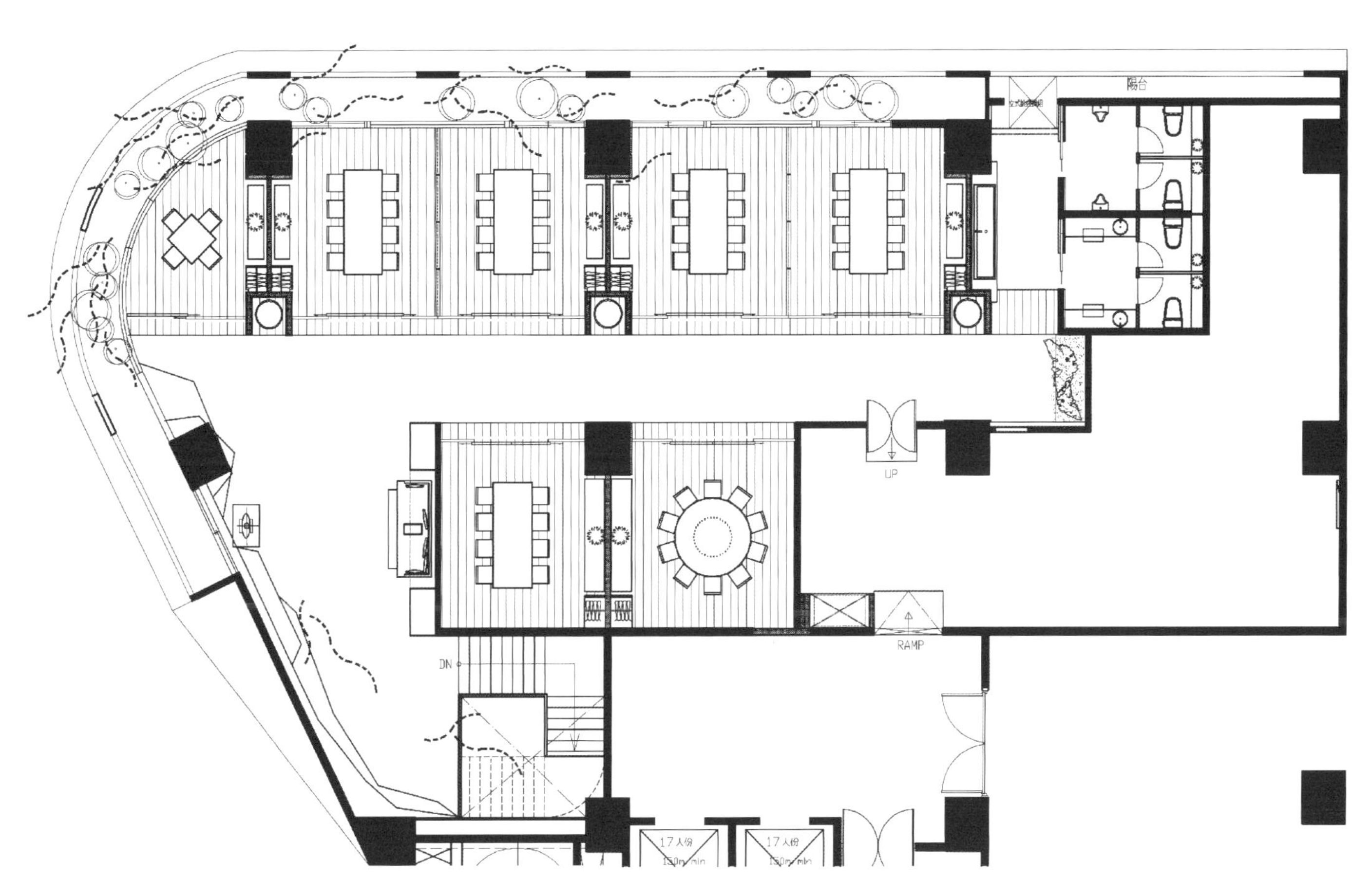

NO. 3 QIAN MEN

前门23号餐厅酒吧

31

【坐落地点】北京前门23号
【面积】2 500 m²
【设计】David Yeo
【摄影】贾方

前门23号酒吧餐厅包括：Shiro Matsu（日式餐厅）、Hex（餐前酒吧）、Agua（西班牙风情餐厅）、Fez（摩洛哥风情酒廊）。

Shiro Matsu是一间设计风格非常现代的日式餐厅，位于具有银座时尚风格的全玻璃建筑的首层。整个内部装饰运用现代艺术理念演绎日本传统竹舍，室内悬挂的盏盏筒形灯和由众多"浮动的蜡烛"组成的"燃烧的岩石"美景，为整个餐厅渲染出一种和谐温馨的气氛。设计师将日式风格设计中的一些精髓元素巧妙穿插于空间中，餐厅中所用的黑色和金色丝绒坐椅面料，由日本知名面料设计师专门度身定做，是独一无二的专属。Hex 酒吧在一层与Shiro Matsu餐厅相连，它包括时尚鸡尾酒吧和酒廊两部分。高高的吧凳、大大的丝绒面料的软沙发、晶莹剔透的鸡尾酒杯，再加上醉人心脾的鸡尾酒，烘托出整个酒吧的格调。

Agua餐厅位于二层，步入餐厅的一瞬间，西班牙的激情洋溢扑面而来，自由奔放的红色布满整个空间，仿佛看见热情的西班牙女郎舞动弗朗明哥，或者英雄的斗牛士在胜利后举杯欢庆。Fez酒吧可谓极具时尚独特之本色，以丰富的元素体现摩洛哥城的浓郁文化风情，灵感来自北京的伊斯兰文化传统风格，该传统在历史上通过著名的"丝绸之路"从中亚引进到中国。

The case of bars and restaurants are divided into: Shiro Matsu (Japanese-style restaurant), Hex (fasting bars), Agua (Spanish style restaurant), Fez (Morocco style lounge).

Shiro Matsu is a very modern style Japanese-style restaurant, located in Ginza fashion style with all-glass building ground floor. The whole interior concept of the use of modern art interpretation of traditional Japanese bamboo homes, indoor hanging lamps and a number of cartridges from the "floating candles," consisting of "burning rocks" beauty for the whole restaurant to evoke a warm and harmonious atmosphere. Designers will be Japanese-style design elements of some of the essence of cleverly interspersed in space, the restaurant used in the black and gold velvet chair fabric, fabric designers from Japan specially made famous, is a unique work. Hex Bar on the first floor connected with the Shiro Matsu Restaurant, which includes the stylish cocktail bar and lounge in two parts. The high bar stool, big soft velvet fabric sofas, crystal clear cocktail cups, plus cocktails, express a whole bar style.

Agua restaurant is located in the second floor, entered the restaurant of the moment, the Spanish passion rushing toward us, free and flowing red filled the entire space, as if to see a Spanish girl dancing Flamingo passion, or the heroic bullfighter in the victory toasting. Fez Bar very stylish features, in order to enrich the element reflects the city's rich culture and customs of Morocco, inspired by Beijing's traditional style of Islamic culture, the tradition's history through the famous "Silk Road" imported from Central Asia to China.

↑满眼的红展现了西班牙的热情 / Eyeful of red to show the Spanish enthusiasm.

↑Shiro Matsu餐厅，地板上镶嵌着一圈通透的玻璃地板 / Shiro Matsu Restaurant, transparent circle on the floor inlaid with glass floors.
↓Hex 餐前酒吧，六边形的造型贯穿空间全部 / Hex pre-dinner bar, hexagonal shape throughout the space for all.

↑玻璃墙体内点缀着粉红色的玫瑰花 / Within the glass walls dotted with pink roses.
↓一层平面图 / 1st floor plan.

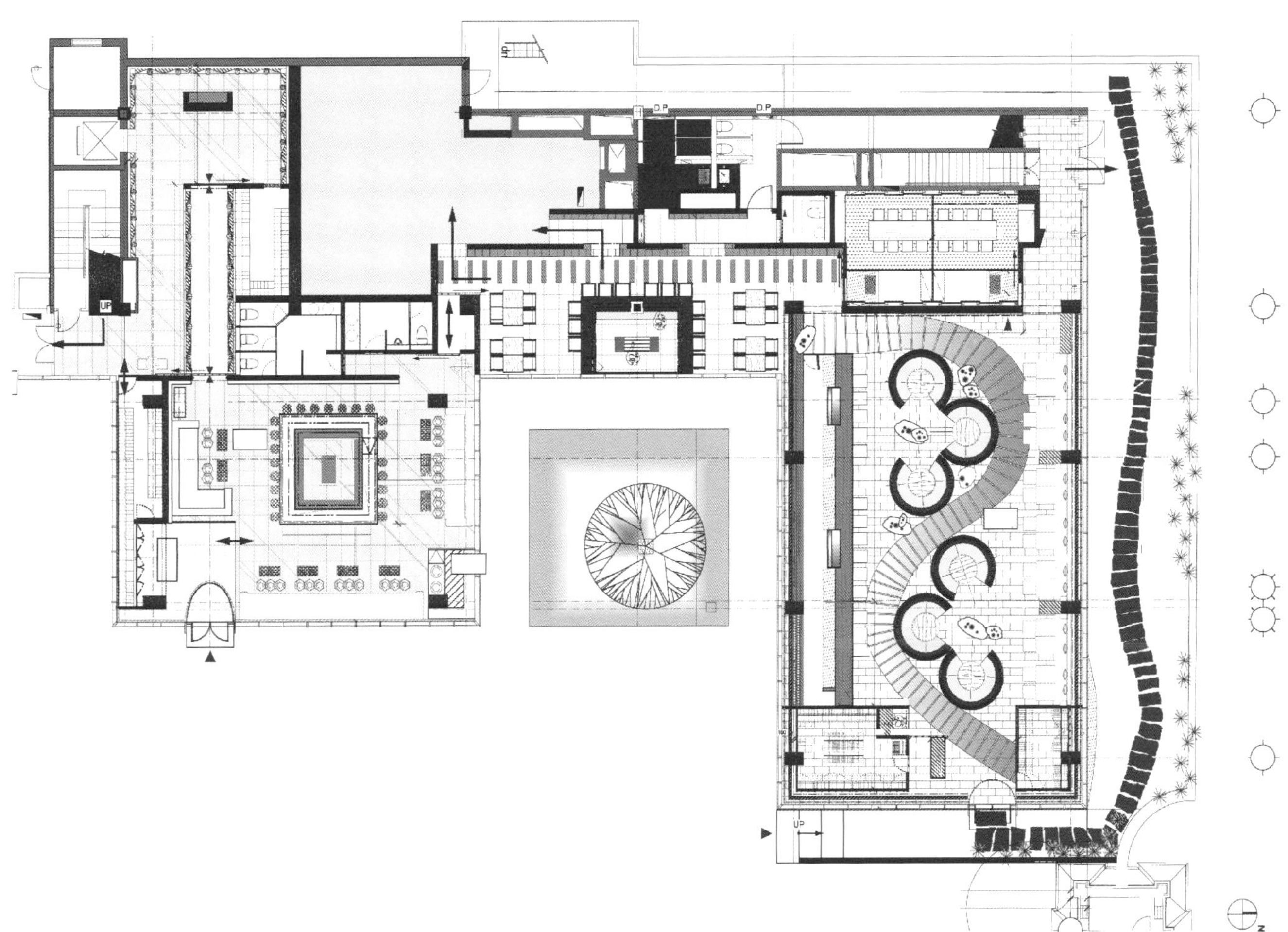

DBL RESTAURANT

DBL餐厅

32

【工程名称】日本DBL餐厅
【面积】211 m^2
【设计】森田恭通〔日〕
【摄影】Seiryo Studio

餐厅直面绿树繁茂的公园，优雅的自然环境成为它天然的前院，也成了设计师的灵感来源。餐厅面积不大，被设计成简单的上下两层式结构。为了让公园内郁郁葱葱的景观在此得到延续，来自新锐漫画艺术家的蔓藤图案艺术品高悬于餐厅之内，与设计师擅长的简约而高雅的设计手法完美结合，相得益彰，为餐厅营造出了既神秘又高雅的设计风格。

DBL餐厅可谓麻雀虽小，五脏俱全，它的内部包含了咖啡厅、餐厅、酒吧三个部分。这个设计以高雅端庄的白色为主色调搭灰色调的黄色和紫色，简单的色彩使用方案突出了餐厅安静优雅的格调。

材料方面，一楼的地面使用了木地板，大部分使用柔软坐面的椅子，点缀宽大的长沙发，简约时尚，二层地面使用了深色的地毯，以宽大的双人沙发为主，感觉也更加豪华舒适。在这个案子中，所有的家具和陈设都是非常简单的，这样就不会和餐厅内复杂的蔓藤花纹图案冲突，张弛有度。

灯光设计是另一大特色。用餐气氛的好坏，除了与餐厅空间的设计和陈设有关之外，灯光更是不容忽视的重要一环。营造一个格调高雅温馨的用餐环境，需要花费大量的心思，只有把握空间照明的原理，精心挑选设计适当的照明设备，悉心研究光线、环境和人的心理的相互关系，才能在这一环节上为设计画上点睛之笔。

Restaurants face the lush green trees of the park, the natural environment into its elegant natural front yard has become a source of inspiration for designers. Dining area is small, is designed to be a simple up and down a two-tier structure. To make the park landscape in this continuity from the comic artists, works of art was placed on the vine pattern within the restaurant, and designers are good at is simple and elegant design the perfect combination of methods, for the restaurant to create a mysterious and elegant design.

DBL restaurant features a very complete, its internal includes cafes, restaurants, bars three parts. The design is elegant and dignified to ride a white-based color gray yellow and purple, a simple color use of the program highlights the quiet and elegant style restaurant.

In the materials, the ground floor using the wood floors, most of the sitting surface of a soft chair, decorative large couch, concise fashion; the second floor on the ground using the dark carpet, to the large double sofa-based, sensory and more luxurious. In this case, all the furniture and furnishings are very simple, so as not to, and restaurant complex conflict vine motifs.

Lighting design is another major feature. Dining atmosphere of good and bad, in addition to dining space with the outside of the design and furnishings, lighting is an important part can not be ignored. To create a warm and elegant dining environment, need to spend a lot of thoughts. Only by grasping the principles of space lighting, carefully selected to design appropriate lighting, careful study of light, the environment and people's psychological relationship between the order in this part of the design to increase strength.

↑室外就餐区 / Outdoor dining areas.

↑灯光点亮的时候，即使是在室外也会被那奇妙的灯光效果吸引 / When the lights lit, even in the outdoors will also be those wonderful lighting effects to attract.
↓餐厅有一个很高的中庭 / Restaurant has a very high atrium.

↑一层的主题墙面图案与中庭悬挂的图案呼应 / The subject of a layer of the atrium wall hanging pattern and echo pattern.
↓一层平面图 / 1st floor plan；↓二层平面图 / 2nd floor plan.

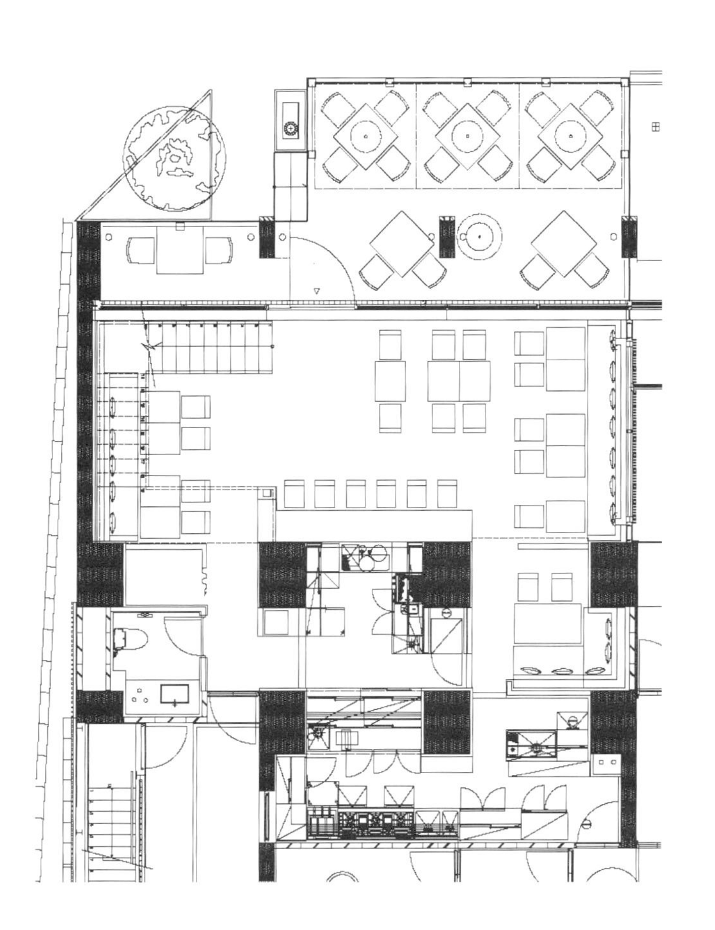

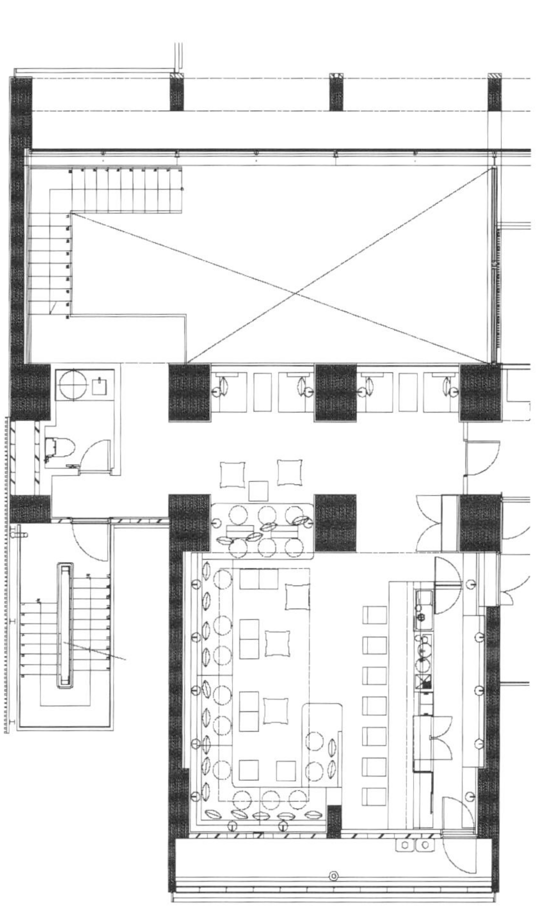

↑↓玻璃墙体内点缀着粉红色的玫瑰花 / Within the glass walls dotted with pink roses.
←高挑的层高 / The storey tall.

THE PERSPECTIVE OF JANPAN

和食莎都上海连锁店

33

【坐落地点】上海长宁区
【工程名称】和食莎都上海连锁店
【面积】435.35 m²
【设计】岩本胜也〔日〕

餐厅的设计任务是和上海已有的众多日本料理店有所区分、融合中国文化、并且为其发展连锁产业做长久的考虑。

餐厅既落户于上海，所以能让上海普通民众接受并喜爱就成了设计工作的重中之重。在提供合理膳食、充实服务的基础之上，设计师更加重视空间自身的易入性和开放感：宽阔无阻的入口是吸引顾客的一个亮点，它几乎毫不费力就将顾客带入了餐厅内部；全场不完全的隔断，让宾客能够在整个餐厅自由游走；顶部整齐排列的白炽灯光带给顾客更加宽广的视觉体验，木质材料在整个空间内贯穿始末，连接着各部分空间。所有这些细节都带给顾客开放、轻松的心情和用餐体验。

和当地文化相融合，在异国餐厅的设计中显得尤为重要，于是"取材中国"也成为了本次设计的重要主题：设计师将丰富的中国素材包含了进来，"四君子"之一的"竹"在这里大行其道，巧妙地将空间隔断，却又留有足够的视觉空间，有种"欲语还休"的魅惑，让顾客在充分享受美食的同时，领略到更深层次的文化融合之美。由于是连锁餐厅，日后连锁店发展的延续性也不可忽视，所以在设计中，"组合式艺术"也成为其中的一个主题：设计师将极富日本风情的剪影用在隔板上，来表现日本四季的变化。

Restaurant design task with the Shanghai's Japanese restaurant has been differentiated, integration of Chinese culture and its development industry to do a long chain of consideration.

Restaurants settled in Shanghai, Shanghai allows ordinary people to accept and love have become the focus of the design work. In the provision of a reasonable diet, on the basis of satisfactory service, designers pay more attention to space itself easy entry and open sense of: broad and unhindered entrance is a focal point to attract customers, which directly within the customer into the restaurant; audience does not completely cut off, so that guests can move freely throughout the restaurant; top of the neatly arranged in incandescent light to bring customers a broader visual experience, wood materials in the whole space throughout the whole story, connecting the various parts of space. All of these details are brought to customers open, relaxed mood and dining experience.

And the local culture integration, the design of the restaurant in a foreign country is particularly important result, "localization" has become an important task of this design: Designers will include a wealth of Chinese material in, "four gentlemen", one of the "Bamboo "Here is extensively used skillfully space partition, yet allow sufficient visual space, a kind of" opaque "the temptation for customers to fully enjoy the food at the same time, to experience a deeper level of cultural integration. Because it is restaurant chain in the future development of the continuity of the chain can not be ignored, so in the design, the "combined art" has become one of the themes: the Japanese-style silhouette design used in the partition on to represent the four seasons in Japan change.

↑细部，非常日本特色的图案 / Detail, very distinctive pattern in Japan.

↑餐厅门口，可以看到内部的通畅 / Restaurant entrance, we can see inside smooth.

↓夹层中的VIP厅 / Dissection in the VIP hall.

↑“竹”，将空间巧妙隔断 / "Bamboo" will space cleverly cut off.
↓平面图 / Plan

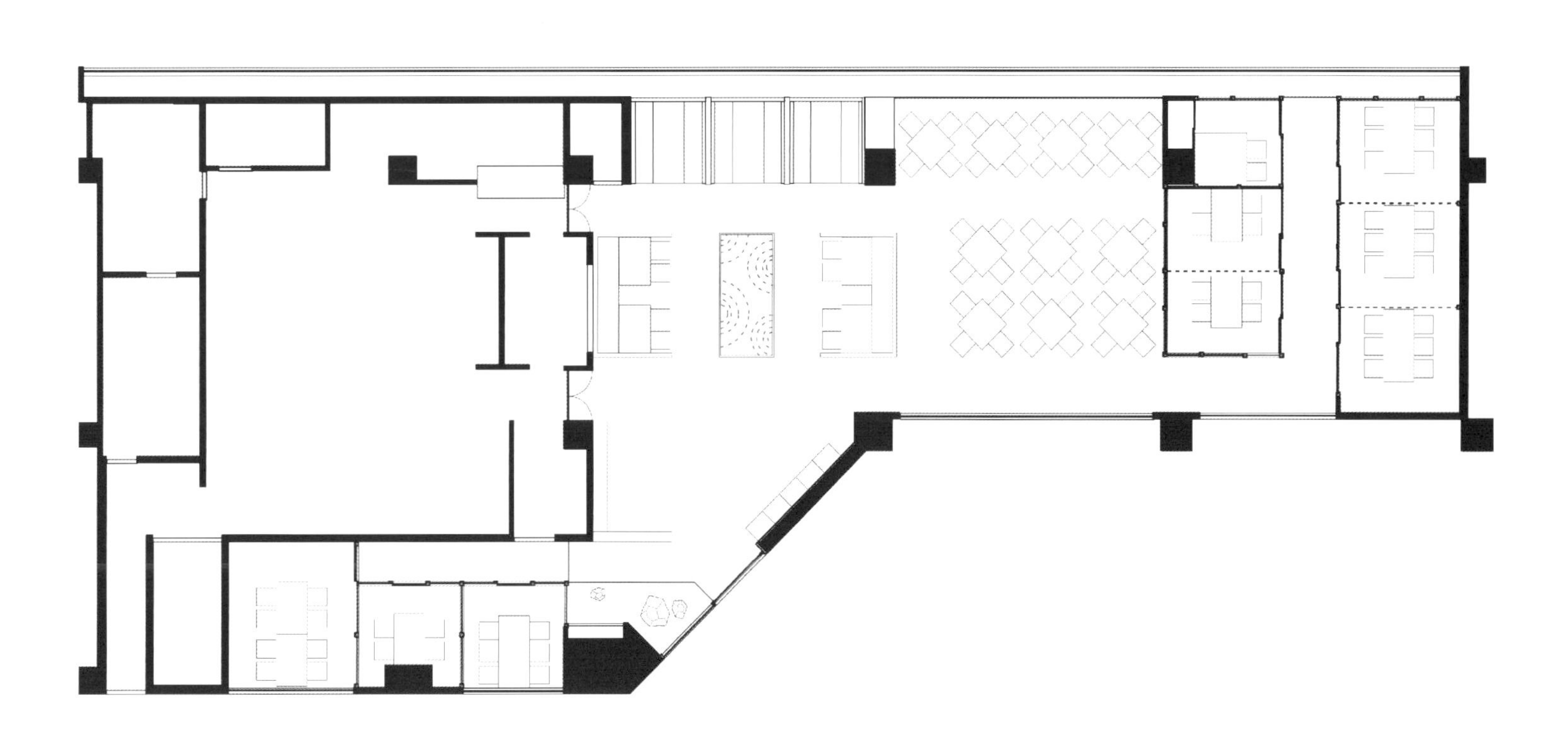

↑纸灯，日本元素之一 / Paper lamps, one of the elements in Japan.
↓宽阔的通道 / Wide channel.

↑特色包间 / Features rooms.
↓竹屏风的细部 / Bamboo screen in detail.

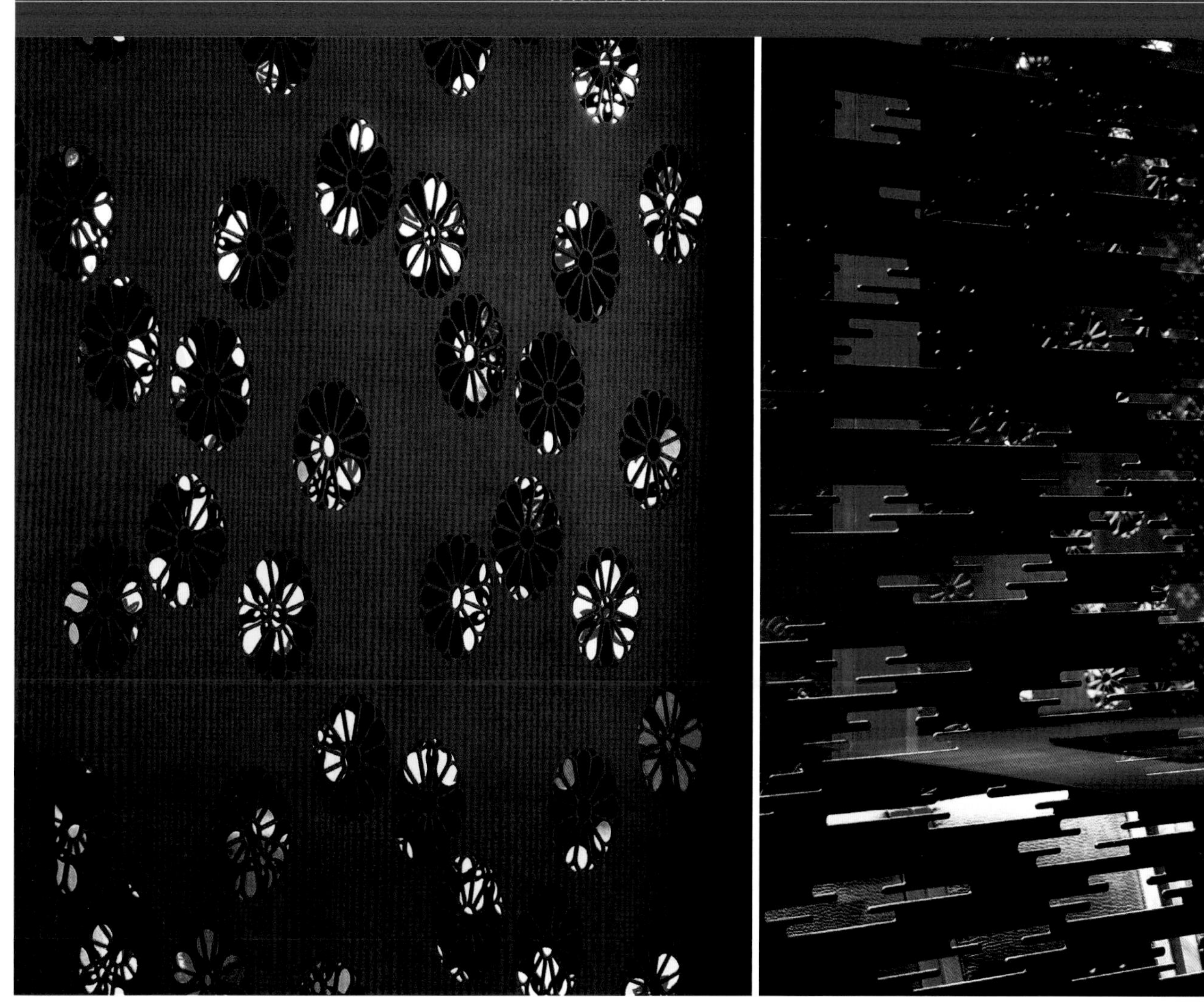

VISUAL FEAST

面加日式拉面馆

34

【坐落地点】上海市静安区愚园路
【面积】270 m^2
【设计】陈幼坚
【摄影】Alvin Chan

设计师意在赋予这个日式拉面馆以现代、年轻和亲切的室内风格，更确切地说应该是一种casual dining风格，就是简洁、不花俏、看似漫不经心，规划却井然有序，高档。品牌希望吸引那些年轻，收入不高，但又愿意花费稍高价位来享受个性餐厅体验的顾客，而这些大都市里的年轻人，大都讲究生活品质，充满活力，并有一双对时尚感非常敏锐的眼睛。于是对于设计师的要求便是要在味觉以外，最大程度的满足这部分人群的视觉需求。

在餐厅主题色的选择上，设计师自然采用了拉面馆标识的颜色：红和白。撇开大面积的白色，我们可以看到，餐厅内的餐垫，部分椅子，通往二楼餐厅的落地玻璃隔断等等都采用了大红色。在吸引人眼球的同时，在中国，红色还是人们最喜欢的颜色之一，有种振奋向上的精神提升和令人食欲大增的奇妙作用。

作为日式拉面馆在餐厅细节上，品牌还是希望增加一些日式的风格。于是特意为餐厅订制的餐椅完美地弥补了这一缺憾。餐椅的设计灵感源自于食用日式拉面时非常典型的汤碗及木筷子的形状。使得顾客在现代、休闲的氛围中用餐时，还能隐约感受到别样的日式风格，非常独特。

Designers intended to give this Japanese-style noodle shop to a modern, young and friendly interior style, but rather should be a casual dining style that is simple, practical, seemingly at random, planning is rigorous, high-grade. Brand to attract the young, income is not high, but higher prices are willing to spend to enjoy the personalized customer dining experience, and these young people in a big city, most pay attention to the quality of life, full of vitality, and a pair of very sharp sense of fashion eyes. So the requirements for the designer is to taste other than the greatest degree of vision to meet the needs of this part of the crowd.

The choice of the theme color in the restaurant, the designers used the restaurant identity of natural colors: red and white. Leaving aside large areas of white, we can see, restaurants inside the mat, some chairs, the restaurant's floor to the second floor of glass partitions, etc., with a large red. In the eye-catching at the same time, the red is the color of the Chinese people's favorite, one of a kind of inspiring spirit and appetite is greatly increased role.

As a Japanese-style ramen museum in a restaurant details, brand, or want to increase the number of Japanese style. So ordered specifically for the restaurant dining perfectly made up for this shortfall. Chair design inspiration comes from eating Japanese-style noodle soup, and when the very typical shape of wooden chopsticks. Makes the customer in a modern, casual dinner atmosphere, but also a vague sense that different kind of Japanese style, is unique.

↑贴有餐厅标志图案的玻璃墙，给人们留下了深刻的印象 / Affixed with the symbol of the glass wall of the restaurant to the people left a deep impression.

↑日本拉面餐厅全貌 / Japanese ramen restaurant picture.

↓鲜蔬照片把健康的餐饮理念传递给顾客 / Fresh vegetables photos healthy food and beverage concept to delivery to customers.

↑餐厅的玻璃墙上的主题图案 / The theme of the restaurant's glass wall pattern.
↓平面图 / Plan

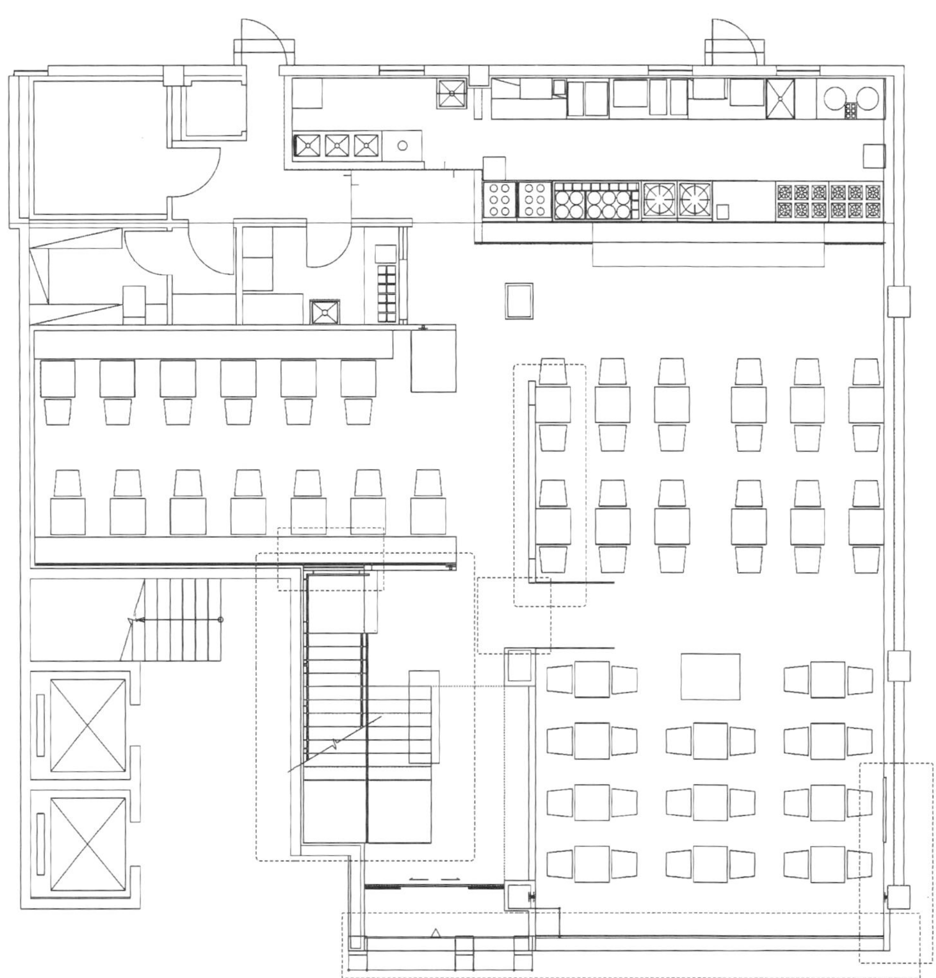

A TAIWAN'S RESTAURANT

山手八番日式料理餐厅

35

【坐落地点】中国台北台中德安购物中心；【面积】300 m²
【设计】朱柏仰；【设计单位】暄品设计
【主要用材】烤漆玻璃、环氧树脂、人造石材、木作喷漆
【摄影】李国民

本案的日式料理餐厅不同于台湾那些贵得离谱的日式料理，它以年轻消费者为目标客群，努力打造亲民的现代日式轻食料理风格。开放式的空间使得顾客可以从商场的其他层就能看到它，两边的长臂部分成为整个设计最难处理的部分，同时也是最需要处理好的部分。

作为设计主轴的折纸造型隔板，由白色喷漆木材和深色的烤漆玻璃就地拼接组合而成，经济简单的选材，并不影响效果的表达。不同造型的折纸造型隔板，既可以是通道隔断，又可以是小包间的隔板，还可以是个性的通道走廊。"折"的立面及线条在狭长的空间里将三度空间的连续性翻转及延展展露无遗，更增加了空间的景深感。

在材料上，设计师紧扣快餐型日式轻食料理的特性，没有使用到任何特殊材料，仅以普通的材料即打造出别样的美感。极具现代感的白色贝壳纹环椅，由塑料和钢材制成，在使用轻巧的同时兼顾到了经济实用。设计师还贴心地考虑很多年轻人可能即将去同层的影院观影或刚从电影院出来，眼睛不适应强灯光，故在灯光的运用上，设计师只用到了小型的轨道灯。轨道灯的设计考虑到折纸造型隔板的设置，同时又与深色的背景墙配合，让顾客有身临电影院之感。

This case is different from the Japanese-style cuisine restaurant in Taiwan that expensive Japanese-style cuisine, it targeted customer group of young consumers, and strive to create cost-effective style of modern Japanese cuisine. Open space allows customers from the mall the other layers will be able to see it, both sides of the long arm of the most difficult part of becoming part of the entire design, but also most in need of a good part of the deal.

Origami spindle shape as a design partition, from white spray paint to paint a dark glass timber and a combination of splicing and economic simple selection does not affect the expression of results. Different forms of origami style partition, both channels can be cut off, but also can be a partition wall between packages, but also can be personalized access corridor. "Discount" in the narrow facade, and lines in the three-dimensional space will be overturned and the extension of the continuity that accentuates, which adds to the depth of field space.

In the material, the designers closely linked to fast food-type Japanese-style features, does not use any special materials, only ordinary materials to create a special feeling. A very modern white shell patterns ring chair, made of plastic and steel in the use of light at the same time takes into account the economic and practical. Designers also intimate account of many young people may be about to go on the same floor of the theater cinema viewing, or just come out, his eyes not suited to strong light, so the use of the lighting, the designers only use a small track lights.

↑空间结构因折纸造型隔板摆放角度的改变而改变 / Spatial structure due to folding partition form laying angle is changed.

↑深色地运用增加空间的纵深 / Dark-depth use of more space.

↓蓝色地使用让餐厅延续了影院的静谧 / Blue continued to use the restaurant theater calm.

↑轨道灯的灯轨在天花板上绘出了一幅几何画 / Track lights on the ceiling light rail plotted a geometric painting.
↓轨道灯的灯轨 / Track lights light rail；↓平面图 / Plan

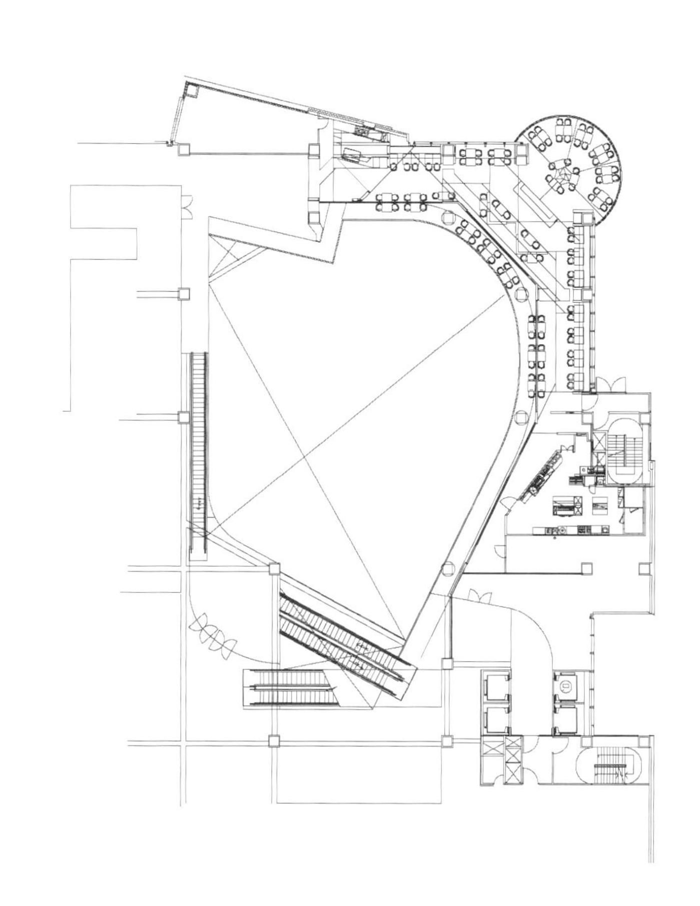

BAR HULU

Bar Hulu餐厅和酒吧

36

【坐落地点】上海中山东一路500号，外滩黄埔公园；【建筑面积】685 m^2

【设计】耿治国；【参与设计】彭兆、许振华；【设计公司】飞形设计事业有限公司

【主要材料】象牙白玉石、镜面不锈钢、陶瓷、皮质铍包

【摄影】周宇贤

本案建筑共有3层，整体呈L形。设计师以"区块"概念对空间进行规划，一楼为接待厅，二楼是员工区和洗手间，三楼是空间主体，主要划分为休闲区和酒吧。

提到Bar Hulu这个名称是由"8"、"吧"、"福禄"、"葫芦"，这四种意象通过谐音和形体的演变完美地融合在一起，物与物、场与场之间产生奇妙的联系，带出一种言已尽而意深远的微妙感觉。

空间场景的塑造是设计的难点，设计师索性抛开所谓"刻意的设计"，以"有序的堆砌"的手法取而代之，在空间中展开了一幅虚拟的立体山水画卷，无论是以包豪斯建筑为设计灵感的带有中国情调的家具，还是在镜面不锈钢上用挤白的工艺再现李可染经典的山水画，无一不是尝试将"东方"概念植入新锐而摩登的娱乐情境中去的空间小品。

设计师在Bar Hulu中引入了纯粹中国的文化元素——山水、松石及饮酒葫芦，这些主题的使用，重新诠释了现代简洁的空间。无处不在的葫芦元素、抽象为极富装饰性的山水纹路，却以富于时代感的新式材质表现，与选用典型中式材质而以西式手段表现的其他装置相辅相成，亦中亦西，巴洛克的东方风就此展现到完美。

Building the case a total of three layers, as a whole was L-shaped. Designers to "block" concept of space planning, on the first floor the reception hall, second floor, the staff area, on the third floor is a space for the main body is divided into the Lounge Area and Bar.

Bar Hulu the name referred to by the "8", "bar", "foo", "gourd", which four kinds of images through the homophonic and physical evolution of the perfect blend together and create wonderful contacts through integration with a wonderful feeling.

Space scene design difficulties in shaping the designers to abandon the "deliberate design", use "orderly design" approach, in space, launched a virtual three-dimensional mountain Chinese painting. Whether in the form of inspiration for the design of Bauhaus architecture with Chinoiserie furniture, or in the mirror stainless steel on the reproduction process with special Li Keran classical Chinese painting, are trying to "Eastern" concept into the cutting-edge entertainment and the modern context work.

Bar Hulu designers introduced purely Chinese cultural elements - landscape, turquoise and drinking gourd. The use of these themes to explain the modern compact space. Gourd decoration elements and landscapes, and lines to enhance the space of oriental exoticism, but the use of Western-style methods, making the performance of the Baroque style of the East even more perfect.

↑镜面不锈钢天花抽象为极富装饰性的山水纹路 / Mirror stainless steel ceiling abstract landscape for the highly decorative lines.

↑接待台，一派中国情调 / Front Desk, school of Chinoiserie.
→吧台前面的装饰采用串联起来的球 / Bar in front of the decorative use of tandem up the ball.
↓无处不在的葫芦元素贯穿空间 / Ubiquitous element throughout the space gourd.

↑窗外景色一览无遗 / View at a glance out the window.
←楼梯间的镜面处理丰富了空间的表现力 / Stairwell enrich the space mirror the performance of processing power.
↓平面图 / Plan

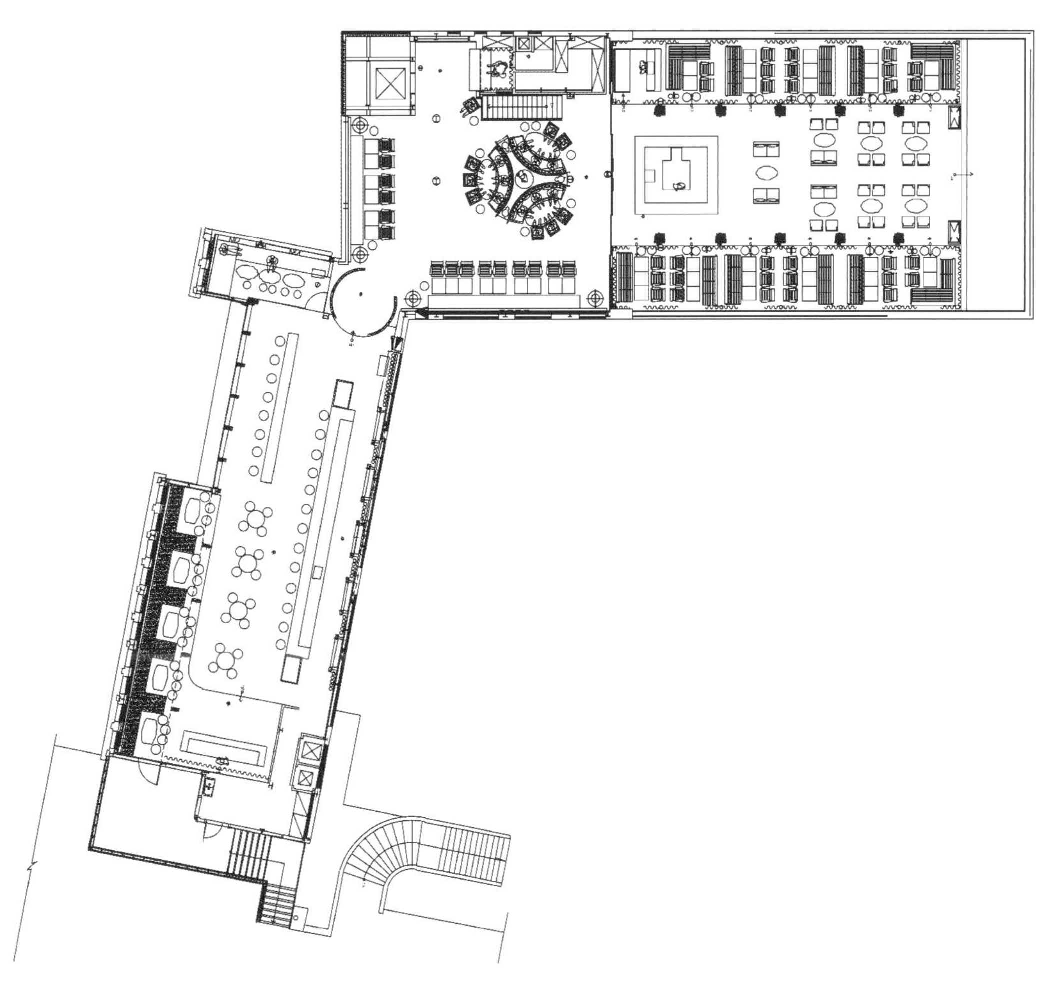

SIN LOUNGE

Sin Lounge酒吧

37

【坐落地点】上海市石门一路
【建筑面积】1 700 m²
【设计】Kokaistudios
【摄影】贾方

酒吧的定位是充满了浪漫主义色彩，安详和激烈、柔和和尖锐，各种能够被用于室内空间的矛盾元素被揉捏在一起——一个汇聚所有美好事物的伊甸园，一个能够让人们的情绪与体验瞬间交融的场所。

从电梯出来就能看到一条长长的走廊，为了满足不同需要，空间被分隔成两个区域。走廊的装饰显然花了不少精力，两边墙面以菱形不锈钢片饰面，着意模仿蛇的表皮纹理。为了达到最佳视觉效果，材料不仅开模定做，而且在结构上也经过了仔细研究。在内部空间，超过7 m的层高让设计夹层成为可能。新建的夹层被用于VIP 私人包间使用，而每个包间都靠窗，以求获得更佳的远眺视野。厨房等辅助设施被安排在夹层的下部，这样安排的好处正在与能保证其他区域的完整，包括服务区及展示区在内，吧台等设施围绕厨房依次展开，最大限度地使空间利用率达到最高。

在诸多元素中，色彩当然是首当其冲的。设计师为不同的VIP房间分别设计了七种不同色彩，紫红色、黄色、橙色、绿色、紫色、大红色及天蓝色。而表现这些色彩则不仅仅是墙面和灯光，包括沙发、窗帘和地毯等软饰在内，都刻意围绕色彩主题挑选，其中一些家具更是经过了特别定制和单独设计。

SIN LOUNGE positioning is full of romantic colors, serene and intense, soft and sharp, can be used in a variety of contradictory elements of interior space has been together——A convergence of all the good things of the Garden of Eden, one can make people's emotions and experience an instant fusion place.

Come out from the elevator you can see a long corridor, in order to meet the different needs of space divided into two regions. Corridor decoration obviously spent a lot of effort on both sides of the wall with diamond-shaped stainless steel sheet finishes, imitating a snake skin texture. In order to achieve the best visual effect, material is not only custom-made mold, but also in structure after a careful study. In the interior space, more than 7 m high layer of the sandwich has been divided out. The new sandwich is used in private VIP rooms, each with private rooms have a window in order to obtain a better field of vision. Kitchen and other ancillary facilities are placed in the lower mezzanine, the benefits of working with such an arrangement can ensure the integrity of other regions, including the service area and display area, including, bar and other facilities around the kitchen, expand, maximizing the space utilization to the highest .

In many elements, the color is the most special. Designer for various VIP rooms were designed in seven different colors, purple, yellow, orange, green, purple, large red and blue. The performance of these colors are not just walls and lighting, including sofas, curtains and carpets, have deliberately chosen the theme around the colors, which is after some of the furniture was custom designed and individual design.

↑水晶吊坠勾勒而成的蛇形吊灯 / Crystal chandelier pendant sketch made of serpentine.

↑Lounge区棱镜饰面扶手围栏 / Lounge Area prism finishes railings.

↓ 入口两侧墙面蛇鳞装饰 / Snake scales on both sides of the entrance wall decoration.

↑Lounge区里如光线流动般的扶手围栏 / Lounge district, such as light flows like railings.
↓一层平面图 / 1st floor plan.

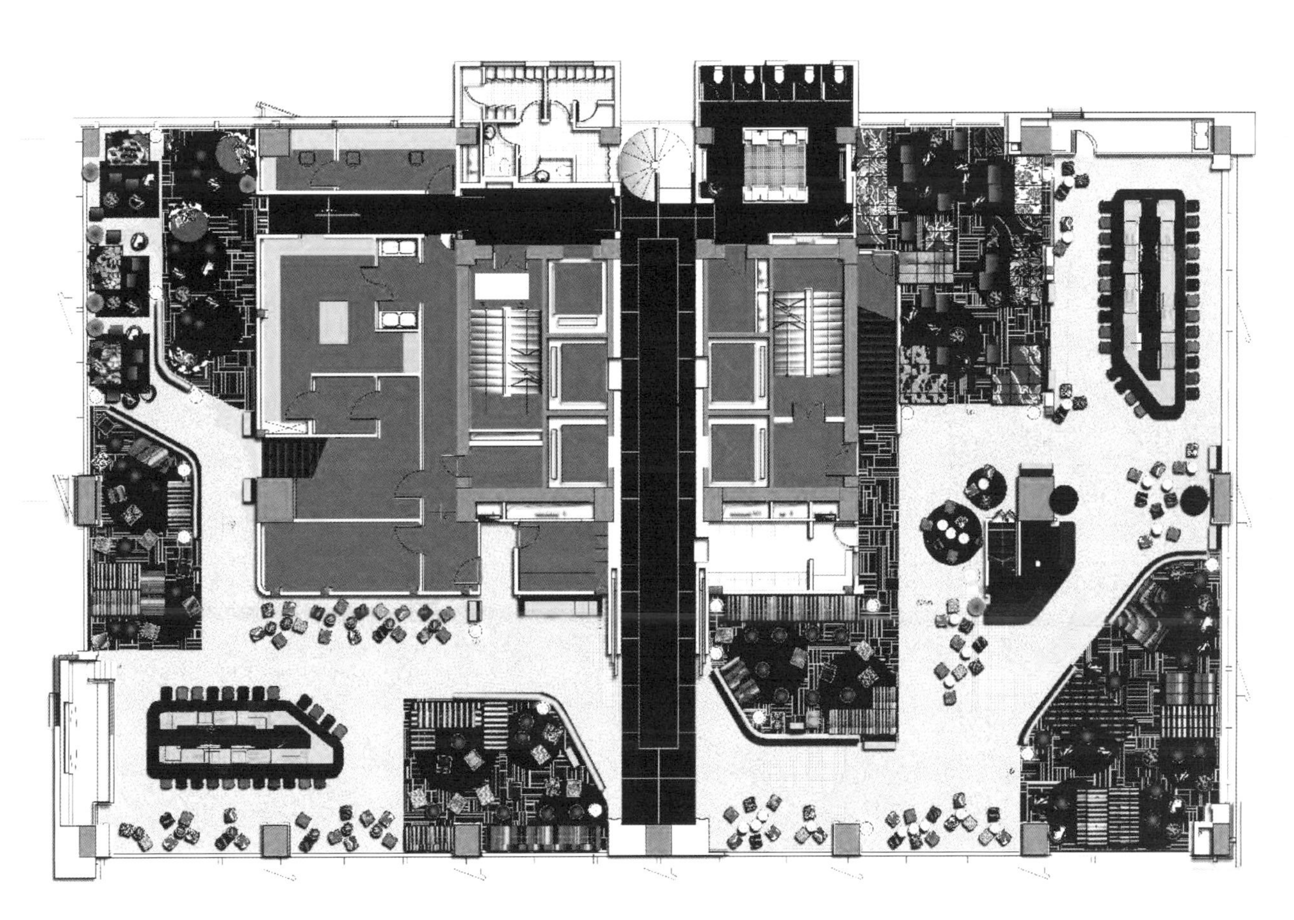

↑以“风”为主题的Disco区 / In order to "wind" as the theme of the Disco zone.
→巨大天使翅膀庇护的DJ台 / A huge angel wings asylum DJ sets.
↓二层VIP包厢区平面 / The second floor VIP balcony area plane.

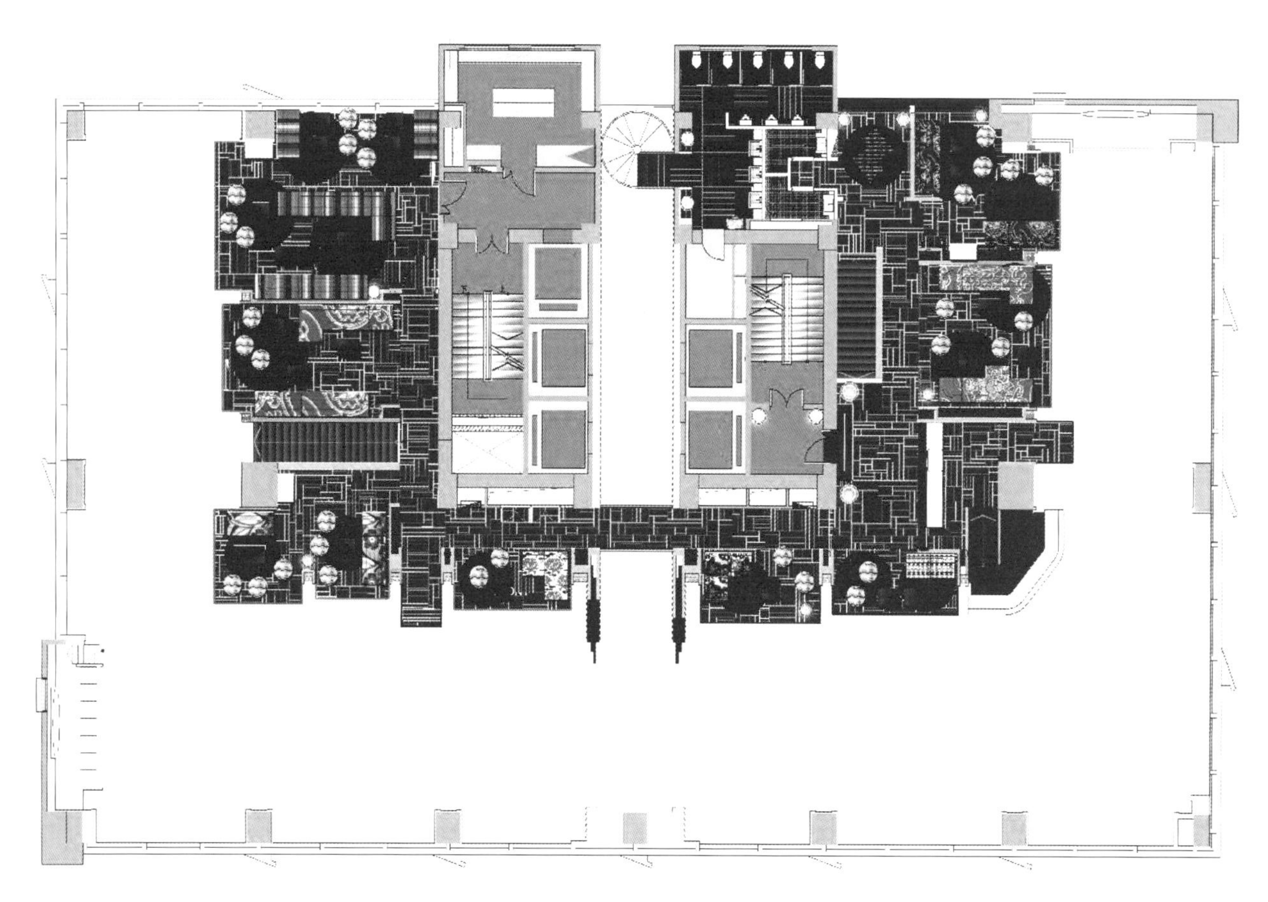

↑↓Disco区内二层各色主题包厢 / Disco theme of the region colored two-story box.
←Disco区一角 / Disco zone corner.

↑半开放的蓝色“夏娃”VIP包厢 / Semi-open blue "Eve" VIP box.
↓红色“亚当”VIP包厢造型别致的沙发 / Red "Adam" VIP box modeling chic sofa.

↑绿色主题的VIP包厢 / Green theme of the VIP box.
↓SIN室内立面图 / SIN interior elevation.

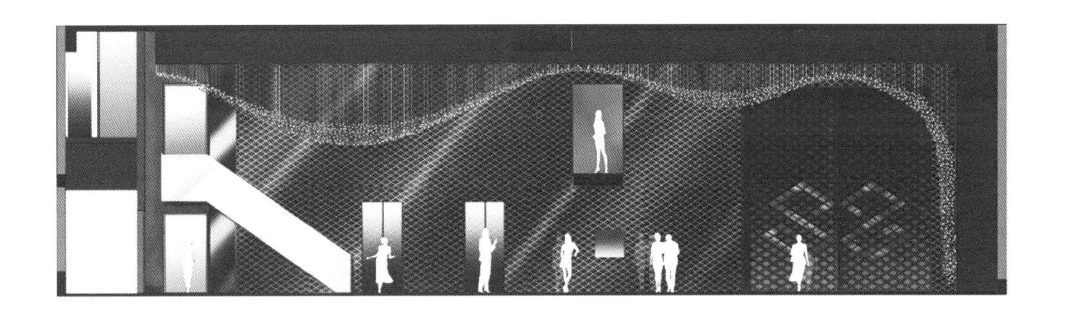

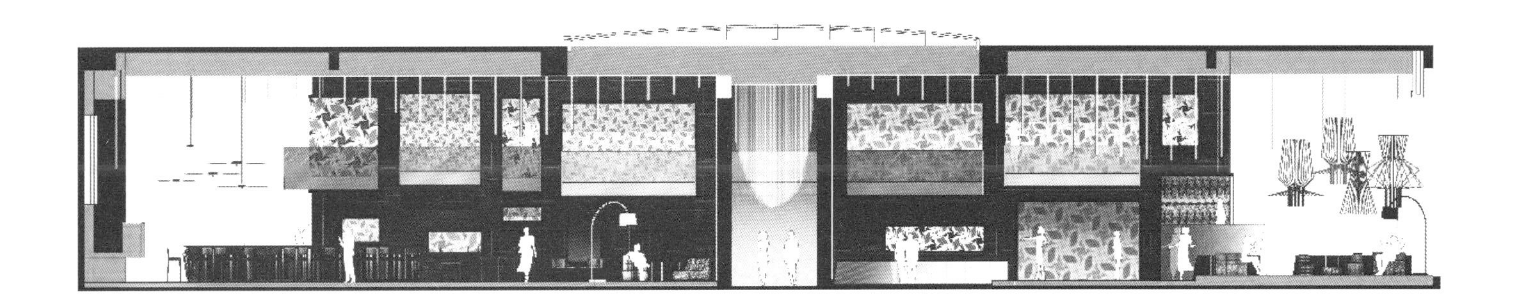

T-O 12 BAR AND CLUB

某酒吧俱乐部

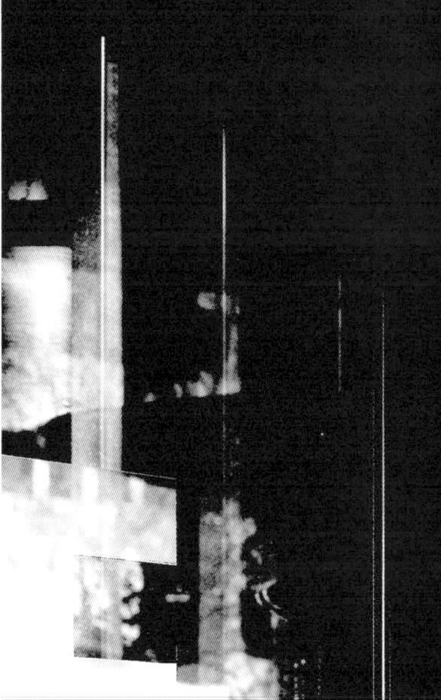

38

【坐落地点】德国斯图加特
【建筑面积】340 m², 1 200 m³
【设计】Ippolito fleitz group〔德〕
【摄影】Zooey Braun

酒吧的定位是充满了浪漫主义色彩，安详和激烈、柔和和尖锐，各种能够被用于室内空间的矛盾元素被揉捏在一起——一个汇聚所有美好事物的伊甸园，一个能够让人们的情绪与体验瞬间交融的场所。

主吧区接待着不同身份的客人，U形吧台由白色可丽耐材料制成，一幅大面积的白色超现实绘画在黑色底纹的墙面上跃然呈现，成为空间中最吸引人的背景。背景图案中还穿插了一些实体绘画元素，从墙面延至空间中。天花板上垂下的镜灯，为空间平添了虚幻高度。底层有两个小型休息区。其中之一的镜厅，房间拥有多边形的镜墙。另一个休息区的空间设计则成为中央桌台的完美陪衬，空间的造型与桌台的形状完全一致。

二层空间由大舞池和长条酒吧间组成。地下一层有小型的舞池和酒吧寄存处。值得一提的是卫生间，同样设计精良的墙面和天花板图案是昆虫和食肉植物的组合，其温暖黑暗的空间中那些小东西随处可见。

Bar determined to make a difference, from design concept to implementation have adopted a new approach. Designers plan to create a conceptual space beyond the nightclub, a strong and unusual visual effects emerged. Bar site is a College of Music, is now divided into three layers. The bottom of the main bar and two small rest area, sub-bar, large dance floor in the second floor, small dance floor was arranged in the ground floor. The space in the store design, designer stores deliberately shrink a length overall inward to give up the aisle space. Entrance is also back in the second floor hallway with the store below the plexiglass box-shaped lights create a long arrogance extends to the second floor door wall, signs on the lights in the entire road is clear, the flow of light to attract the customers.

The main bar area with different identities reception guests, U-shaped bar by a white Corian material, a large area of white surrealistic painting the wall black shading rendering, becoming the most attractive space background. Background images are also interspersed a number of entities, elements of the painting from the wall until the air. Mirror hanging ceiling lights, adding to the illusion of space height. There are two small ground floor lounge area. Among them, one is composed of a mirror hall, the room has a polygon mirror wall. Another rest area, space design has become the perfect foil for the middle desk sets, desk sets the shape of the shape and the same.

The second floor space by a large dance floor and long bar of the composition. Basement has a small dance floor and bar storage. It is worth mentioning that the bathroom, the same well-designed walls and ceiling designs are a combination of insects and carnivorous plants, its warm, dark space, the little things everywhere.

↑巨大的黑白图案，妖娆的女郎散发着青春的热情 / A huge black and white pattern, enchanting young girl exudes youthful enthusiasm.

↑↑一幅大面积的白色超现实绘画展现了生活的乐趣 / A large area of white paint to show the joy of life.
←一层主吧区，U形吧台 / A layer of the main bar area, U-shaped bar.
↓酒吧入口立面图 / Bar entrance elevation.

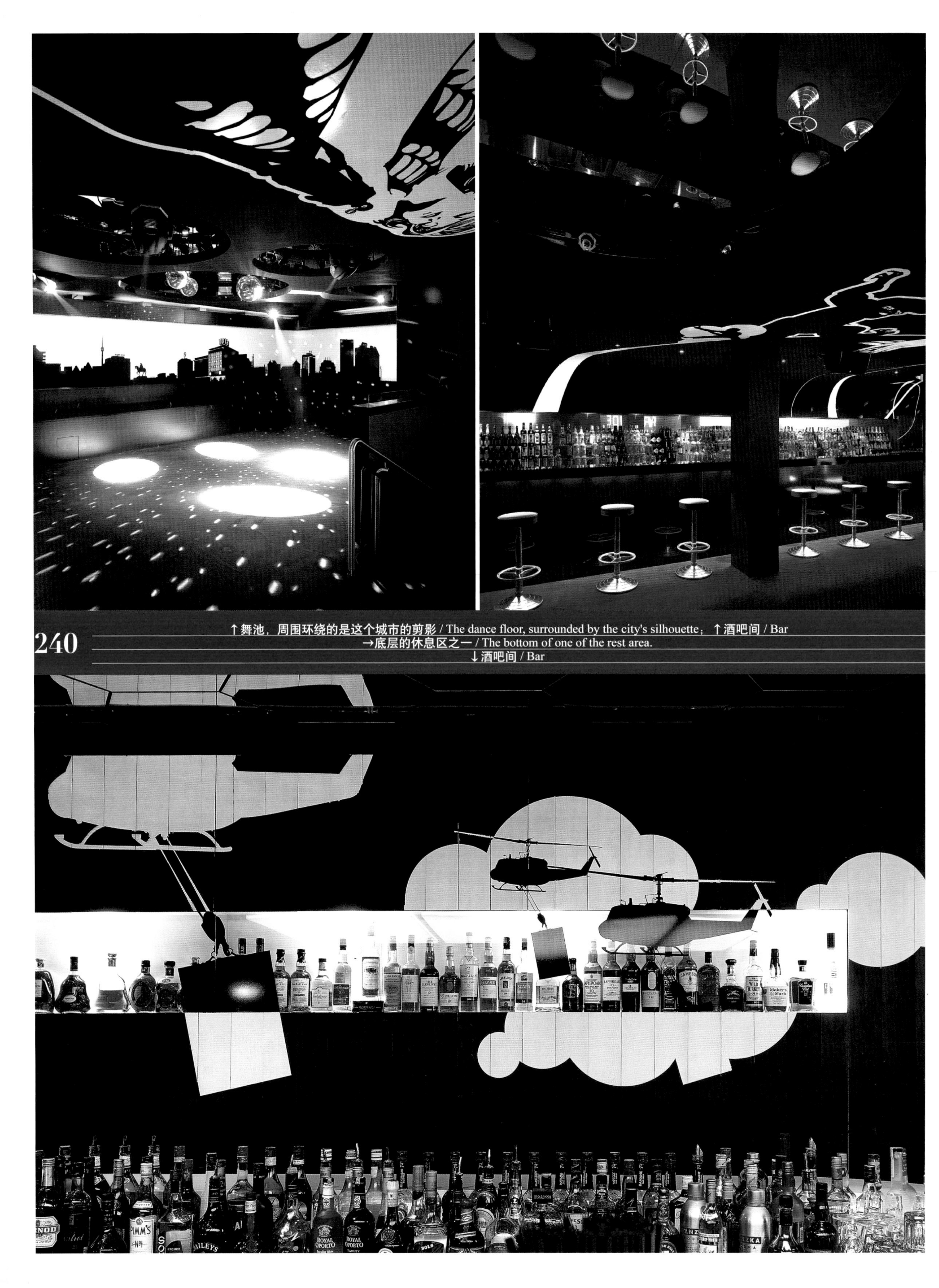

↑舞池，周围环绕的是这个城市的剪影 / The dance floor, surrounded by the city's silhouette；↑酒吧间 / Bar
→底层的休息区之一 / The bottom of one of the rest area.
↓酒吧间 / Bar

↑↑楼梯转角处 / Corner stairs.
←图案的变化始终是T-O 12的特色 / Pattern of change has been the T-O 12 features.
↓楼梯间 / Stairways；↓卫生间 / Toilet

THE WALL BAR IN SHANGHAI

上海The Wall Bar

39

【坐落地点】上海浦东滨江大道2727号广场地下一层；【建筑面积】980 m^2
【设计】胡汉淞；【参与设计】丛夕铭、庄铁佳
【设计单位】上海元典建筑设计咨询有限公司
【摄影】王瑞璠

上海THE WALL的前身是一条狭长幽暗的地下通道，这里被我们视为是一堵蕴藏着这个大都市的年轻活力、热烈期待着被发掘及自我释放的"空间墙"。

为了强化这个狭长空间的特殊魅力及透视空间的视觉震撼力，我们提出了"空间中的空间"、"场景中的场景"的概念，试图在这单一狭长的大空间场景中，串联交织出不同意涵及动能的次空间。

在这里，有交流互动的场所，有独坐小饮、消磨时光的吧台，有高潮迭起的DJ舞池，还有逃避生活压力的鸦片床……各种瞬间交替重叠，孕育出充满都市活力与诱惑的迷人空间。

在这个透视感极强的狭长空间中，从左到右串联交织着四个主题空间场景："迷宫"、"戏揄"、"爱神"和"逃避"。这四个主题隐含着四个不同历史时代的含义：从秦陵俑坑的神秘、西潮东渐的奢华、浪漫诱惑的情爱，到远离现实生活、自我销蚀的鸦片时代。

Shanghai THE WALL, formerly known as a narrow dark tunnel, here is that we considered to be a wall bears the city's youthful vigor, warm look forward to being discovered and self-release of "space walls."

In order to strengthen the special charm of this narrow space and the visual perspective of space power to shock, we have proposed the "space in space" and "Scene of the scene" concept, trying to narrow the large space of this single scene, creating a different connotation and dynamic sub-space.

Here, there is place for interaction, there sat a small drink, spend time bar, there is one climax after another of the DJ dance floor, as well as to escape the pressure of life of opium bed ... ... a variety of moments alternately overlapping, and carries out full of urban vitality and temptation the charming space.

In this perspective a very strong sense of narrow space, from left to right there are four themes space scene: "labyrinth", "drama draw out", "Eros" and "escape." These four themes implicit in the meaning of four different historical era; from the ancient oriental mystery, a Western luxury, romantic lure of love, to stay away from real life, self-decadent era of opium.

↑ 大空间中的小场景 / Large space in a small scene.

↑蓝色的LED灯布 / The blue LED light cloth.
↓手绘图 / Hand drawing.

↑蓝色的LED灯布 / The blue LED light cloth.
↓平面图 / Plan

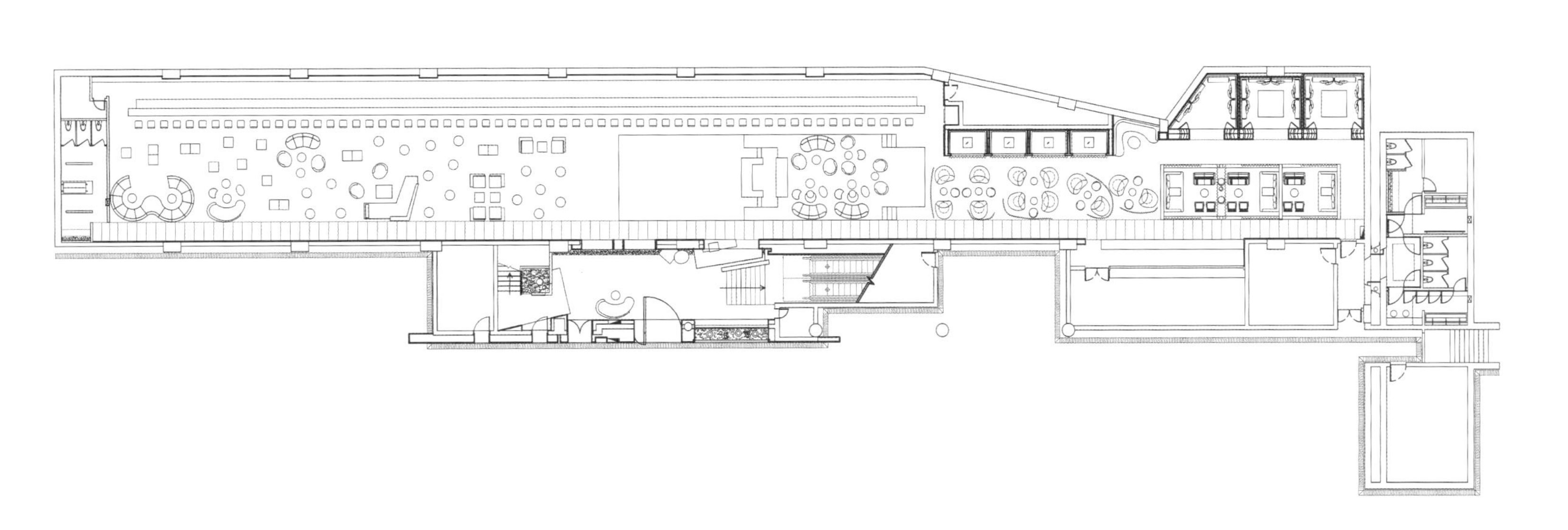

↑旧式的家具营造形成一个华丽的小空间 / The formation of the old furniture to create a beautiful small space.
←纱幔中的小包厢 / Gauze in a small box.
↓手绘图 / Hand drawing.

LIP BAR IN SHEN ZHEN

深圳“唇”酒吧

40

【坐落地点】深圳福田区；【建筑面积】230 m^2
【设计】据宾
【主要用材】GRG，木材，地砖
【摄影】据宾、方振鹏

酒吧“唇”的设计构想完全来自于设计师的主观与随性。说主观，是因为承接该项目的前提是设计师需要绝对的主导权，包括酒吧的类型及风格都要由设计师来决定；说随性，是因为设计的过程是天马行空的，完全没有刻意而为。

“唇”处于一个大型商场的一层，位于深圳市最中心的繁华地段，周围均是充斥着世界名牌的大型购物中心（shopping mall）。在通常的印象中，这类区域无一不是灯红酒绿、纸醉金迷，满是大都市特有的喧嚣与迷乱。而设计师将“唇”定位在“静吧”，就有反其道而行之的想法。满眼各种形状的白色，已经为这个空间定下了基本的调子。

一层是散座区，中心吧台的设计似动还无，让人无法知晓下一秒它将变幻出何等景象，而镂空的花墙则让室内与室外有了对比与交流。二层有三个包厢，弧形的长洞是包厢与外界最直接的联系，极具现代主义的兴味。因为想体现光，所以用了LED。而这个使用是极为夸张和非凡的。在白色的大背景下，暧昧的红、忧郁的蓝、温暖的黄、宁静的绿，色彩肆意而纯粹地变幻。

Bar "lip" design concept entirely from the designer's subjective and with the nature. That subjective, because to undertake the project on the premise that designers need absolute leadership, including the type and style of the bar should be decided by the designer; argued that with, it is because the process of designing an exercise in imagination, there is no deliberate and for.

"Lip" in a layer of large shopping malls, located in Shenzhen, most central downtown location, surrounded by are filled with world famous shopping malls. In the usual impression, such has been a bustling area full of city-specific noise and confusion. The designers of "lip" positioning in the "quiet," is there to find quiet in noisy thoughts. Eyeful of various shapes white, has been set for this space the basic tone.

Layer is scattered seat area, center bar design are constantly changing, people do not know how in an instant it changes out of sight, while the hollow of the wall displaying the indoor and outdoor has been compared to the exchanges. The second floor there are three boxes, a long curved hole box the most direct contact with the outside world, very interesting modernism. Because you want to reflect the light, so use the LED. And this use is extremely exaggerated and extraordinary. In the white background, ambiguous red, melancholy blue, warm yellow, serene green, the color changes arbitrarily and purely.

↑中心吧台似一泓流动的池水 / Center bar like flowing water.

↑↑GRG与LED的完美融合 / GRG perfect integration with the LED.
↓二层包厢过道 / The second floor balcony hallway.

↑圆台极具现代主义的意味 / Round very modernist edge；↑散座区一角 / Corner of seat area.
↓一层平面图 / 1st floor plan；↓二层平面图 / 2nd Floor Plan

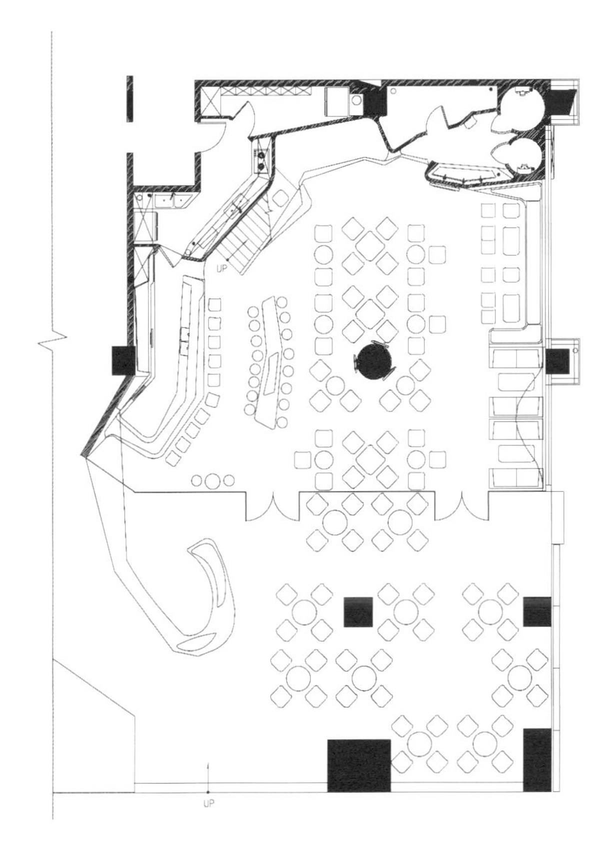

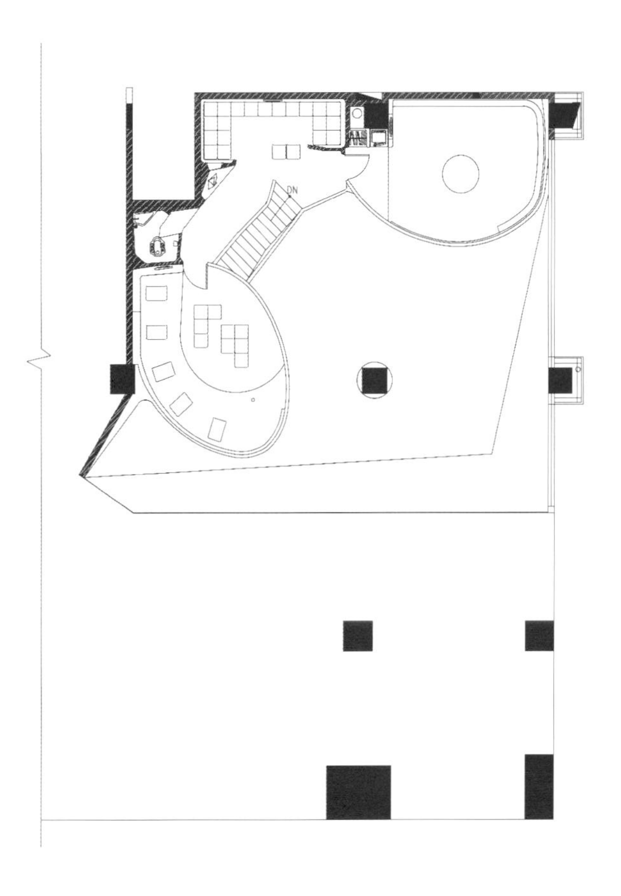

↑↑包厢内的光影变幻 / Changes in light and shadow box
←从二层俯视 / From the second floor overlooking.
↓↓包厢内的光影变幻 / Changes in light and shadow box.